Aspects of Illegal, Unreported and Unregulated Fishing in the Southern Ocean

Reviews: Methods and Technologies in Fish Biology and Fisheries

VOLUME 5

Series editor:

Jennifer L. Nielsen

U.S. Geological Survey,
Biological Resources Division,
Anchorage, Alaska

Aspects of Illegal, Unreported and Unregulated Fishing in the Southern Ocean

by

Rachel J. Baird
University of Queensland,
Brisbane, QLD, Australia

Springer

A C.I.P. Catalogue record for this book is available from the Library of Congress.

ISBN-10 1-4020-5338-X (HB)
ISBN-13 978-1-4020-5338-2 (HB)
ISBN-10 1-4020-5339-8 (e-book)
ISBN-13 978-1-4020-5339-9 (e-book)

Published by Springer,
P.O. Box 17, 3300 AA Dordrecht, The Netherlands.

www.springer.com

Cover photo: Dissostichus eleginoides (Patagonian Toothfish)

Printed on acid-free paper

Wild is the cry of the sea in the caves by it-
Sea that is smitten by the spears of the snow;
Desolate songs are the songs of the waves by it-
Down in the south where the ships never go.

(Extract from Beyond Kerguelen by Henry Kendall. H. Kendall, Leaves from an Australian Forest-Poetical Works of Henry Kendall (1869). This poem was written following a whaling trip in 1853 to the Southern Ocean by 14 year old Henry Kendall.

Table of Contents

Abstract

Overfishing threatens the viability of high seas living resources. Furthermore, controls to prevent overfishing are inadequate. Illegal, Unreported and Unregulated (IUU) fishing is a product of overfishing and affects all marine fisheries. Southern Ocean fisheries have been particularly targeted by IUU fishing. The efforts of the *CCAMLR* Commission and the Australian government to adopt complementary measures to deter IUU fishing in the Southern Ocean form the basis of this study, although the examination is conducted following a comprehensive review of the legal and other factors contributing to the development of IUU fishing as well as the applicable international law.

The conclusions reached in the final Chapter are that in the absence of flag State control, alternative measures and strategies have proved to be effective in influencing the behaviour of IUU fishing vessels. By improving coastal State surveillance, enhancing regional cooperation, imposing port and market State controls, establishing IUU vessel databases and vigorously prosecuting offenders, RFMOs and coastal States can jointly increase the risk of conducting IUU fishing activities. At the same time financial returns can be diminished so that engaging in IUU fishing becomes economically unattractive. It is concluded that the application of a combination of these measures is required in order to achieve a degree of success in deterring IUU operators.

This study is up to date in law and facts to 30 October 2005.

Abbreviations

Abbreviation	Full Name
AAD	Australian Antarctic Division
AFZ	Australian Fishing Zone
CCAMLR	Convention on the Conservation of Antarctic Marine Living Resources
CDS	Catch Documentation System
Code of Conduct	Code of Conduct for Responsible Fisheries 1995
Compliance Agreement	Agreement to Promote Compliance with International Conservation and Management Measures by Fishing Vessels on the High Seas 1993
Conservation Convention	1958 Convention on Fishing and Conservation of Living Resources of the High Seas
EC/EU	European Community/European Union
EEZ	Exclusive Economic Zone
CITES	Convention on International Trade in Endangered Species 1976
FAO	Food and Agriculture Organisation
GATT	General Agreement on Tariffs and Trade 1947
ICJ	International Court of Justice
ILC	International Law Commission
IPOA-IUU	International Plan of Action- Illegal, Unreported and Unregulated Fishing 2001
IMO	International Maritime Organisation

ISOFISH	International Southern Ocean Long line Fisheries Information Clearing House
ITLOS	International Tribunal for the Law of the Sea
IUU Fishing	Illegal, Unreported and Unregulated Fishing
IWC	International Whaling Commission
LOSC	Law of the Sea Convention 1982
NAFO	Northwest Atlantic Fisheries Organisation
NGO	Non-Governmental Organisation
OECD	Organisation for Economic Cooperation and Development
RFMO	Regional Fisheries Management Organisation
UN	United Nations
1992 UNCED	United Nations Conference on the Environment and Development 1992
UNCLOS III	Third United Nations Convention on the Law of the Sea
UN Fish Stocks Agreement	United Nations Agreement for the Implementation of the Provisions of the United Nations Convention on the Law of the Sea of 10 December 1982 relating to the Conservation and Management of Straddling Fish Stocks and Highly Migratory Fish Stocks 1995.
USA	United States of America
USSR	Union of Soviet Socialist Republics
VMS	Vessel Monitoring System
WTO	World Trade Organisation

Tables, Graphs and Maps

PART 1

The Problem of IUU Fishing

to marine fishing activities in all sections of the world's oceans. Chapter 1 also provides a synopsis of the international response to IUU fishing over the past decade, proceeds to an overview of the extent of IUU fishing, both as a global problem and as experienced regionally in the Southern Ocean. Finally it introduces the question of why IUU fishing has been able to develop and persist, a question that must be considered before proceeding to identify measures to address the growing and persistent problems created by IUU fishing. This question is discussed more comprehensively in Part 2.

The analysis in Part 2 of the legal and non-legal factors contributing to the development and persistence of IUU fishing is fundamental to making an informed review of the actions of *CCAMLR* and Australia. Significant progress in identifying factors has been undertaken in the last two years, particularly following the conclusion of the OECD *Workshop on IUU Fishing Activities* held in April 2004.[3] After reflecting on the 'drivers' of IUU Fishing, this text proceeds to examine the substance of the several international instruments with application to IUU fishing with the aim of identifying areas of emergent State agreement or practice and clarifying the international framework supporting the implementation of legal measures to deter IUU fishing in the absence of effective flag State enforcement.

Parts 4 and 5 analyse the extent to which *CCAMLR* and Australia have implemented the measures identified in Part 3 on a regional and national level, and the degree of success experienced by both *CCAMLR* and Australia in deterring IUU fishing. The final Chapter of this text draws from the analysis in Parts 2-5, to identify and recommend those measures which can potentially assist in deterring IUU fishing. The concluding Chapter concludes that first, no one measure can be relied upon to deter IUU fishing, and second, that to be effective, measures must be formulated by taking account the factors identified in Part 2 as contributing to the development and persistence of IUU fishing.

3 *OECD Workshop on IUU Fishing Activities*, Paris, 19-20 May 2004. <www.oecd.org/document/5/0,2340,en_2649_33901_21007109_1_1_1_1,00.html> (12 July 2004).

Chapter 1

The Impact of IUU Fishing on Marine Fisheries

1.1 The Physical and legal State of Marine Fisheries

1.1.1 Overfishing

The existence of overfishing was recognised by the first FAO Fisheries Technical Committee in 1946.[4] The recorded global marine capture fisheries catch increased by as much as 6% per year during the two decades following 1950.[5] In simple terms

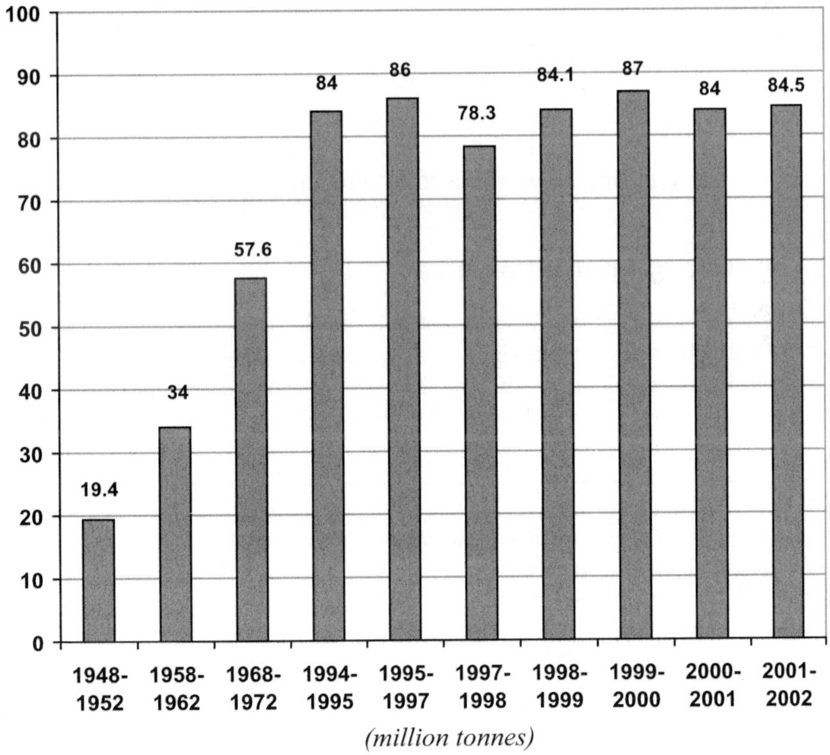

Graph 1.1 Global Marine Catch in 1948-2002[8]

4 FAO, *The State of World Fisheries and Aquaculture Report* (2004) Part 2, 88.

5 FAO, *The State of World Fisheries and Aquaculture Report* (2000) Part 1, 1.

the reported catch trebled from 18 million tonnes in 1950 to 56 million tonnes in 1969.[6] Whilst increases were not as significant during the last thirty years of the 20th century, the global landings from marine capture fisheries continued to rise. The most recent FAO estimate places marine fisheries production for 2002 at 84.5 million tonnes.[7]

The continued increase in landings has corresponded with an increase in the proportion of estimated overexploited and depleted fish stocks. In 1949 a review of marine resources charted those areas of the world's oceans considered to contain unexploited resources. The area charted covered most of the ocean's surface.[9] Nearly thirty years later in 1997, the FAO estimated that up to 60% of marine fisheries were over exploited, fully to heavily exploited or depleted. A further 25% of fisheries fell within the moderately exploited category, whilst only 9% were considered under exploited.[10] In 2003 FAO estimates placed up to 76% of stocks within the over exploited, fully to heavily exploited, depleted or recovering categories. It was estimated that just 3% of fish stocks remained under exploited.[11]

It is acknowledged that approximately 90% of the global marine catch is taken from within coastal State declared EEZs,[12] however this study does not seek to specifically examine the management (or mis-management) of stocks falling within EEZs, with the exception that the review of Australia's efforts to deter and eliminate illegal fishing within the declared EEZ adjacent to the Heard and McDonald Islands necessitates such a review. Rather, the complex problem of IUU fishing, which by definition can occur in disparate maritime zones, forms the central theme of this study.[13]

1.1.2 The Role of International Law

The absence of a single regulatory authority with respect to the high seas has meant that many high seas stocks (including highly migratory and straddling stocks) are particularly vulnerable to overfishing. The principle of freedom of fishing on the high seas, which derives its authority from the observance of a general high seas

---

6 Ibid.

7 FAO, *State of World Fisheries and Aquaculture Report* (2004) Part 1, 4.

8 Information for Graph 1.1 was obtained from FAO, *The State of World Fisheries and Aquaculture Reports* (1995) (1997) (2000) and (2004).

9 D.H. Cushing, *The Provident Sea* (1988) 258, 296.

10 L.W.Botsford, J. C. Castills and C.H. Peterson, 'The Management of Fisheries and Marine Ecosystems' (1997) 277 (5325) *Science* 509, 510. See also FAO, *State of World Fisheries and Aquaculture Report* (1997).

11 FAO, *State of World Fisheries and Aquaculture Report* (2004) Part 1, 32. The term 'fully exploited' was utilised in terms of a fishery producing catches close to maximum sustainable limits.

12 Division for Ocean Affairs and the Law of the Sea (DOLAS), Office of Legal Affairs, United Nations, *The United Nations Convention on the Law of the Sea- A Historical Perspective* (1998) (copy on file). See also, R. Churchill and A. Lowe, *The Law of the Sea* (1988 2nd ed) 232.

13 The term 'IUU fishing' is explored and defined further is this Chapter.

freedom, means that in practice there is open access to high seas fishery resources for all States.[14] In the absence of a regulatory authority, flag States have been entrusted with the responsibility of exercising jurisdiction over their flagged vessels on the high seas[15] and with ensuring that the general limitations on the exercise of the freedom of fishing, contained within Arts 117- 119 of the *LOSC* are observed.[16]

It should be understood that in instances where flag States are either unwilling or unable to exercise effective control over their flagged fishing vessels, the existing law of the sea, with the exception of specific provisions in the *UN Fish Stocks Agreement*, does not permit third State enforcement.[17] There is a growing body of work addressing the scope for non-flag State enforcement in high seas fisheries[18] however this particular issue is not within the focus of the author's study and it will not be explored further.

---

14 *United Nations Convention on the Law of the Sea,* opened for signature 10 December 1982, 19 ILM 1261 (entered into force on 16 November 1994) ('*LOSC*'). Articles 87 and 116. It is acknowledged that there remains conflict between the observance of the Grotian concept of the freedom of the highs seas (discussed in detail in Chapter 2) and the concept of the common heritage of mankind through which common resources would be collectively managed to avoid the environmental destruction often realised through unregulated access to natural resources. For commentary on this conflict see generally Chapters 1 and 2 of D. Caron and H. Scheiber, (eds) *Bringing New Law to Ocean Waters* (2004). This thesis is based on the realities of overfishing and increased competition for marine fishery resources which are a consequence of the freedom of fishing and which contribute to the current State of IUU fishing. See Part 2 for the analysis of factors contributing to the existence of IUU fishing.

15 The International Law Commission (ILC) emphasised the role of flag states in maintaining order on the high seas in the Report of the ILC on the work of its seventh session 23 April-4 July 1956, *Yearbook of the ILC* (1956) Vol. II. 279. The need for order on the high seas and the role of flag States in exercising authority over their vessels is emphasised further in Part 2.

16 As of 11 March 2004 there were 148 documents of ratification or accession lodged with the Secretary General of the United Nations under Art 306 of the *LOSC*. It argued that the general obligation to cooperate with respect to the conservation and management of high seas resources through subregional or regional fisheries organisations is an obligation under customary international law. This point is examined and substantiated in Part 3.

Indeed the view of the FAO is that the provisions apply to all states including non-member states to the *LOSC* as the requirements of Art 116-119 reflect existing customary international law. See, FAO *State of World Fisheries and Aquaculture Report* (2004) Part 1, Box 9, 98.

17 Several RFMOs have adopted Observation and Inspection Schemes which allow at-sea boarding, however they are only applicable between member States. *The United Nations Agreement for the Implementation of the Provisions of the United Nations Convention on the Law of the Sea of 10 December 1982 relating to the Conservation and Management of Straddling Fish Stocks and Highly Migratory Fish Stocks,* opened for signature 4 December 1995, 34 ILM 1542 (entered into force 11 December 2001) ('*UN Fish Stocks Agreement*') does envisage boarding and inspection by non-flag States. However it is also restricted to State parties to the *UN Fish Stocks Agreement*, the significant change being that it seeks to widen boarding and inspection powers through reliance on membership of the *UN Fish Stocks Agreement* and not membership of the relevant RFMO. Closer inspection shows it remains dependent on State consent to the proposed boarding and inspection regimes.

18 For further analysis of the issue of non-flag State enforcement, see R. Rayfuse, *Non-Flag State Enforcement in High Seas Fisheries* (2004) and also Rayfuse, 'Countermeasures and High Seas Fisheries Enforcement' (2004) 51 *Netherlands International Law Review* 41-76.

The resultant lacuna in international law whereby fishers can operate in an environment lacking effective regulation in which RFMOs and coastal States (beyond their EEZs) have no direct enforcement powers to prevent non-member States or third States from fishing unsustainably, has contributed to the growth of IUU fishing.[19] The legal limitations characterise IUU fishing as a low risk, high return undertaking and large corporate entities, attracted by the economic benefits to be gained with minimum legal risk, have recently entered the IUU fishing industry. The impact of IUU fishing in the Southern Ocean has been particularly severe. The remote location of fishing grounds and absence of any permanent coastal State presence[20] make it relatively easy for IUU fishing vessels to avoid detection, further tilting the odds in favour of the IUU fisher or operator.[21]

There is also a myriad of other factors contributing to the continued rise in global landings of marine fish stocks and the emergence of IUU fishing. These include the impact of the industrialisation of the fishing industry; increases in both human consumption and the size of the global fishing fleet; the introduction of government subsidies which have contributed to the creation of an artificial environment of profitability;[22] increased competition amongst fishers and the entry of large scale commercial fishing entities into the marine fishing industry.[23]

---

[19] Indeed the 1999 Report of the Secretary General to the 53rd Session of the General Assembly of the United Nations on Oceans and law of the sea (A/54/429) 30 September 1999, agreed as much. At paragraph 249 it is states: 'IUU fishing also raises some fundamental issues associated with well-established norms and principles of international law relating to the qualified freedom of high seas fishing, a flag State's exclusive jurisdiction over vessels flying its flag on the high seas, rules regarding treaties and third states and the duty to cooperate for the conservation and management of living marine resources of the high seas.' ('1999 Report of UNSG on Oceans and law of the sea')

[20] Although the French do maintain a scientific base at Port-aux-Francais on Kerguelen Island, there is no permanent enforcement capability on the sub-Antarctic islands. French surveillance patrols have traditionally sailed from la Reunion, a French Island in the Indian Ocean and Australian vessels have left from the port of Fremantle in Western Australia. Coastal State surveillance and enforcement activity is reviewed in Part 5.

[21] The term IUU operator is used on occasion through the study to acknowledge that there is an entity behind the IUU fisher, more often than not directing the location and fishing intensity of the vessel. The operator may be a corporation or individuals.

[22] C.C. Schmidt, 'Addressing Illegal, Unreported and Unregulated (IUU) Fishing' (paper prepared for the *OECD Workshop on IUU Fishing*, above) 14. Schmidt makes a reference to the 2002 Declaration of the World Summit on Sustainable Development to 'put into effect the FAO international plans of action to prevent, deter and eliminate illegal, unreported and unregulated fishing by 2004' and 'eliminate subsidies that contribute to illegal, unreported and unregulated fishing and over-capacity.' The impact of subsidies on high seas fishing is examined in Part 2.

[23] Some of these factors are listed in D. Doulman, 'Illegal, Unreported and Unregulated Fishing: Mandate for an International Plan of Action' (paper prepared for Expert Consultation on IUU fishing organised by the Government of Australia in cooperation with the FAO, Sydney 15-19 May 2000) http://www.affa.gov.au/corporate_docs/publications/pdf/fisheries/eciouuf/ausiuu20003.pdf (12 May 2001). A copy of this paper and other papers prepared for the Expert Consultation can be found in FAO, *Report of and Papers presented at the Expert Consultation on Illegal, Unreported and Unregulated Fishing* FAO Fisheries Report No. 666 (2001). See for example, paragraph 6 where the author states: 'Other considerations contribute significantly to IUU fishing. These issues which

The degree to which IUU fishing impacts on international, regional and national efforts to sustainably manage dwindling high seas fish stocks, particularly in the light of continued reported global landings, and the urgent need for a co-ordinated response to the problem has been emphasised by the statement released by the Australian Government in March 2005:

> IUU fishing on the high seas is a highly organised, mobile and elusive activity undermining the efforts of responsible countries to sustainably manage their fish resources. International cooperation is vital to effectively combat this serious problem. By using regional fisheries management organisations as a vehicle for cooperation, fishing states, both flag states and port states, and all major market states, should be able to coordinate actions to effectively deal with IUU fishing activity.[24]

1.2 Understanding IUU Fishing

The definitions of the nature and scope of IUU fishing outlined in the *IPOA-IUU* are adopted.[25] Illegal fishing means fishing within a declared EEZ without the permission of the relevant coastal State or fishing within an RFMO area of application by a vessel flagged to a contracting party.[26] It also encompasses fishing in violation of national or international obligations. Unregulated fishing includes fishing within a RFMO area of application by a vessel either without nationality or flagged to a non-contracting party and which is either inconsistent with, or contravenes the conservation and management measures of the relevant RFMO. It also includes fishing on the high seas in the absence of flag State authorisation. Unreported fishing includes misreporting catch levels or failing to report catches.[27]

---

are often economically motivated, include the existence of excess fleet capacity, the payment of government subsidies (particularly where subsidies maintain or increase capacity) and ineffective fleet monitoring, control and surveillance'. The factors contributing to both the development and persistence of IUU fishing are reviewed in Part 2.

24 Australian Department of Agriculture, Fisheries and Forestry, 'Overview of Illegal Unreported and Unregulated Fishing' <www.affa.gov.au/content/print.cfm?objectid+446232C4-D2A2-4D87-9FE743D> (16 March 2005).

25 *International Plan of Action to Prevent, Deter and Eliminate Illegal, Unreported and Unregulated Fishing* (*IPOA-IUU*) adopted by the 25th session of the FAO Committee on Fisheries on 2 March 2001. Section II, 3.

26 Under Article 21(1) of the *UN Fish Stocks Agreement*, a State which is party to the *UN Fish Stocks Agreement* but not a member of a relevant RFMO may still find its flagged vessels legitimately boarded and inspected by a member state of that RFMO, on the basis of consent obtained by virtue of ratification of the *UN Fish Stocks Agreement*. This Art also implies compliance by the *UN Fish Stocks Agreement* State party regardless of membership of the RFMO. See, also Article 8(3). On this basis, fishing within a RFMO area of application by a UNFSA contracting party could also fall within the definition of IUU fishing.

27 *IPOA-IUU*, Section II(3).

The full text of the relevant section of the *IPOA-IUU* is reproduced below. Whilst illegal and unreported fishing is by definition wrongful, unregulated fishing may, in some circumstances, be legal under international law. That is, all illegal fishing is by necessity unregulated but not all unregulated or unreported fishing, however undesirable or troublesome it may be to RFMOs and coastal States, is necessarily illegal. The FAO Technical Guidelines for Responsible Fishing (No. 9) *Implementation of the IPOA-IUU* note that unregulated fishing that is 'likely to frustrate the achievement of sustainable fisheries' would fall within the scope of the *IPOA-IUU*[28] and as such RFMOs and States should take action to deter fishers from engaging is such fishing.

Any discussion of IUU fishing and the characteristics of IUU fishers, IUU vessels or the flag States that facilitate IUU fishing by failing to exercise the required level of control, will encounter concepts of good and bad actors.[29] In addressing the many issues raised by IUU fishing and identifying alternate legal measures, this study does not seek to impose such labels on States or individual actors. However, the demonstrated behaviour of some States and IUU operators may necessarily lead one to conclude that their actions are undesirable and counter-productive to the goal of sustainable fisheries management. Such a conclusion is quite distinct from stating that the States, corporate entities or individuals involved in IUU fishing are 'bad actors' within the framework of international fisheries law.

[28] FAO Technical Guidelines for Responsible Fishing No. 9. *Implementation of the IPOA-IUU* (2002), 6.

[29] For example see the paper by D. Balton, 'Dealing with the "Bad Actors" of Ocean Fisheries' (paper prepared for the OECD *Workshop on IUU Fishing Activities,* above) Also published in OECD, *Fish Piracy: Combating Illegal, Unreported and Unregulated Fishing* (2004) 57-65.

IPOA-IUU
II. Nature and Scope of IUU Fishing and the International Plan of Action

3. In this document:

3.1 Illegal fishing refers to activities:

3.1.1 conducted by national or foreign vessels in waters under the jurisdiction of a State, without the permission of that State, or in contravention of its laws and regulations;

3.1.2 conducted by vessels flying the flag of States that are parties to a relevant regional fisheries management organization but operate in contravention of the conservation and management measures adopted by that organization and by which the States are bound, or relevant provisions of the applicable international law; or

3.1.3 in violation of national laws or international obligations, including those undertaken by cooperating States to a relevant regional fisheries management organization.

3.2 Unreported fishing refers to fishing activities:

3.2.1 which have not been reported, or have been misreported, to the relevant national authority, in contravention of national laws and regulations; or

3.2.2 undertaken in the area of competence of a relevant regional fisheries management organization which have not been reported or have been misreported, in contravention of the reporting procedures of that organization.

3.3 Unregulated fishing refers to fishing activities:

3.3.1 in the area of application of a relevant regional fisheries management organization that are conducted by vessels without nationality, or by those flying the flag of a State not party to that organization, or by a fishing entity, in a manner that is not consistent with or contravenes the conservation and management measures of that organization; or

3.3.2 in areas or for fish stocks in relation to which there are no applicable conservation or management measures and where such fishing activities are conducted in a manner inconsistent with State responsibilities for the conservation of living marine resources under international law.

3.4 Notwithstanding paragraph 3.3, certain unregulated fishing may take place in a manner which is not in violation of applicable international law, and may not require the application of measures envisaged under the International Plan of Action (IPOA).

The term IUU fishing is a recent addition to the vernacular of international fisheries management. The *CCAMLR* Commission established under *CCAMLR* is credited with being the first RFMO to formally recognise the problem of non-compliant fishers and to coin the phrase 'IUU fishing'.[30] The 1997 Report of the Standing Committee on Observation and Inspection noted that vessels flagged to *CCAMLR* members had been observed fishing illegally within *CCAMLR* waters[31] and the EEZs adjacent to the Prince Edward Islands (South Africa), Crozet Island and the Kerguelen Islands (France), and the Heard and McDonald Islands (Australia).[32] The Committee also noted the increasing incidence of fishing within the Convention Area by non-contracting party States. This activity was classified as 'unreported and unregulated fishing by non-Members'.[33]

Whilst the distinction was made by the *CCAMLR* Standing Committee on Observation and Inspection and the *CCAMLR* Commission between illegal fishing on the one hand and unreported and unregulated fishing on the other, the adoption of the term 'IUU fishing' by the international community has partially blurred the distinction. This is because IUU fishing has been used as a generic description of fishing activity which 'undermines efforts to conserve and manage fish stocks in all capture fisheries.'[34] Further, whilst the definition of IUU fishing within the *IPOA-IUU* has been described as 'comprehensive and systematic and in that sense useful'[35] it does merge distinct activities performed in distinct maritime jurisdictions.

It is true that some IUU fishing activities may overlap. For example, illegal fishing is by its very nature also unreported fishing. However, it is also important to recognise that the generic term IUU fishing is often used to describe specific situations such as fishing within coastal State waters without coastal State permission, or fishing within RFMO waters by a flag vessel of a contracting party in contravention of the relevant RFMO conservation measures. In this context the term illegal fishing would be more accurate than IUU fishing to describe the factual situation. For the purposes of this study the acronym IUU fishing is used generally to refer to the global issues raised.[36] When discussing discrete issues such as foreign vessels fishing without authorisation

30 *CCAMLR*-XVI (1997) Report of the Sixteenth Meeting of the Commission, paragraphs 8.7-8.13, *CCAMLR*-XVI (1997) Report of Standing Committee on Observation and Inspection, paragraphs 1.4-1.22.

31 *CCAMLR*, Article 1, defines the Convention Area in terms of its application to 'Antarctic marine living resources of the area south of 60 degrees south latitude and the Antarctic marine living resources of the area between that latitude and the Antarctic Convergence which form part of the Antarctic marine ecosystem.'

32 *CCAMLR*-XVI (1997), paragraph 1.5.

33 Ibid paragraph 1.20.

34 *IPOA-IUU*, Section I, Introduction.

35 M. Hayashi, 'Illegal, Unreported and Unregulated (IUU) Fishing: Global and Regional Responses' in D. Caron and H. Scheiber (eds) *Bringing New Law to Ocean Waters* (2004) 95, 98.

36 In this context it is noted that the European Community has stated that the definition of IUU fishing in paragraph 3 of the *IPOA-IUU* is not entirely appropriate but could be accepted in the interest of supporting the adoption of the *IPOA-IUU* with the understanding that the EC would not recognise

within the Australian EEZ adjacent to the Heard and McDonald Islands, the term illegal fishing is also used. Fishing within the *CCAMLR* Area of Application[37] by vessels flagged to contracting parties or acceding states, in contravention of *CCAMLR* conservation measures, is strictly illegal fishing. However, given that the problem is not confined to contracting parties, as many non-contracting parties fish the Southern Ocean, the generic term IUU fishing has been adopted by the *CCAMLR* Commission when referring to problems of unsustainable and non-compliant fishing.

The difficulties in employing a broad term like IUU fishing to refer to all forms of unsustainable and non-compliant fishing whether in coastal State waters, RFMO areas of application, or the high seas, is highlighted by the care that must be taken in identifying effective strategies for its deterrence. Whilst this study does not approach IUU fishing in terms of identifying discrete legal measures for each of the constituent activities (that is illegal fishing, unregulated fishing and unreported fishing) it does of necessity, review those measures more appropriate for implementation on a regional basis and those which are appropriate for coastal State implementation. It does this through the examination of *CCAMLR* and Australian responses to IUU fishing.

1.3 Global Attention on the problem of IUU fishing

IUU fishing is a serious global problem.[38] It has been the subject of numerous conferences and in 2001 the FAO adopted the IPOA-IUU. In late 2003 a High Seas Task Force was formed following the June 2003 meeting of the OECD Round Table on Sustainable Development.[39] The Task Force is, as defined by its title, one concerned with high seas fisheries with a stated mission to identify global, regional

---

this definition as having any force other than in the context of the *IPOA-IUU*. W. Edeson, 'The International Plan of Action on Illegal, Unreported and Unregulated Fishing: The Legal Context of a Non-Legally Binding Instrument' (2001) 16 *International Journal of Marine and Coastal Law* 603, 619. Edeson further states that 'it is probable that if the term IUU fishing is used in other contexts, it will now be assumed to be referring to the definition found in the *IPOA-IUU*... despite its imperfections.'

37 *CCAMLR*, Article1.

38 In the 2000 Report on the work of the open ended Informal Consultative Process on Oceans and the law of the sea, it was noted that the prevalence of IUU fishing was considered to be one of the most severe problems currently affecting world fisheries. UN Doc A/55/247 (2000) Part B, paragraph 16.

39 See <http://www.high-seas.org/>. The High Seas Task Force was formally launched on 1 December 2003. See also OECD, 'Ministerial Task Force to Tackle Illegal, Unregulated and Unreported Fishing' News Release 1 December 2003. <www.oecd.orf/documents/51/0,2340,en_1649_201185_20897011_119690_1_1_1> (7 April 2004). Members are the United Kingdom (Chair), Australia, Canada, Chile, Namibia and New Zealand.

The fact that the Round Table on Sustainable Development became involved with the problem of IUU Fishing indicates not only the global nature of the problem, but also the limited success international and regional initiatives and instruments were achieving in the effort to deter IUU fishing. See Submission by the High Seas Task Force to the *International Conference on the Governance of High Seas Fisheries and the UN Fish Agreement* 1-5 May 2005, 3. <http://www.high-seas.org/> (11 May 2005).

and national initiatives to minimise incentives to carry out IUU fishing on the high seas.[40] However it is logical that the problem of IUU fishing within EEZs is not excluded from the Task Force's considerations. The Ministerial Task Force met for the first time in early March 2005.[41] Ministerial members agreed upon six priority action areas for the coming year. These are sharing intelligence on IUU vessels, developing a global register of high seas fishing vessels, preparing guidelines on the performance of flag states regarding their high seas vessels, strengthening in-port measures and control over nationals, analysing trade-related measures and addressing RFMO based initiatives and governance issues.[42]

In April 2004 the OECD Committee on Fisheries hosted a *Workshop on IUU Fishing Activities* with the aim of examining the problem of IUU fishing from a 'multi-disciplinary approach.'[43] In particular, economic and social 'drivers' of IUU fishing were examined by Workshop participants.[44] IUU Fishing was once again the focus of international attention during the *International Conference on the Governance of High Seas Fisheries and the UN Fish Stocks Agreement – Moving from Words to Action* held in May 2005. The High Seas Task Force, represented by Ministers from Canada, Australia and Namibia, re-visited the persisting themes of IUU fishing in its submission to the Conference. Factors identified as contributing to the development and persistence of IUU fishing and targeted areas of action formed the substance of debate at these recent international gatherings and the outcomes are examined in Part 2 and 3.[45] First, however, it is useful to review the international response to IUU fishing over the last decade.

1.4 Gaining Momentum in the International Response to IUU Fishing

The persistent and global nature of IUU fishing was noted in a statement made at the close of the FAO's 25th Committee on Fisheries (COFI) in March 2003. It was

---

40 High Seas Task Force home page, ibid.

41 See HSTF03-HSTF10- Documents relating to the First meeting of the Task Force. <http://www.high-seas.org/> (11 April 2004).

42 See *First Meeting of the High Seas Task Force: Summary of Outcomes* (HSTF/10) 14 March 2005. <http://www.high-seas.org/> (14 May 2005). See also, Fisheries and Oceans Canada, 'Illegal, Unreported and Unregulated (IUU) Fishing and International Fisheries Governance, Key Topics at International Meetings' Media Release 11 March 2005. <www.dfo-mpo.gc.ca/media/newsrel/2005/hq-ac08_e.htm> (20 March 2004).

43 The objective of Part 2 of this study is to examine factors (legal and non-legal) which have contributed in the past or continue to contribute to the creation of an environmental (legal, economic, political) which is conducive to IUU fishing activities.

44 See *OECD Workshop on IUU Fishing Activities* above. Some Workshop papers have been published in hard copy by the OECD in a publication titled: *Fish Piracy: Combating Illegal, Unregulated and Unreported Fishing* (2004) OECD.

45 St John's, Newfoundland and Labrador, Canada. 1-5 May 2005. See Conference home page at <http://www.dfo-mpo.gc.ca/fgc-cgp/index_e.htm> (14 May 2005).

acknowledged by the members of COFI that 'IUU fishing and its impact on resource sustainability is a matter of international concern'.[46] The impact of reflagging on the conduct of high seas fisheries, an activity closely associated with IUU fishing, was recognised much earlier than 2003.[47] The following review of the international awakening (and response to) the growing problem of IUU fishing and its affiliated activities, provides a necessary foundation for this study and in particular for the examination in Part 3 of the international declarations, plans, codes, documents and agreements aimed at deterring and eliminating the practice.

1.4.1 A Blueprint for Action

In 1992 at the UN Conference on the Environment and Development, (UNCED) the international community adopted within Agenda 21, a blueprint for the protection, use and development of the ocean's living resources.[48] Problems of unregulated fishing and reflagging were recognised as contributing to the inadequate management of high seas fisheries.[49] One of the priorities for State action was the need, 'consistent with international law, to deter reflagging of vessels by their nationals as a means of avoiding compliance with applicable conservation and management rules for fishing activities on the high seas.'[50] A few months prior to the UNCED, the Declaration of Cancun was adopted at the International Conference on Responsible Fishing. Participating States at the conference agreed and declared that:

> States should cooperate on bilateral, regional and multilateral levels to establish, reinforce and implement effective means and mechanisms to ensure responsible fishing on the high seas, in accordance with the relevant provisions of UNCLOS.

and

> States should take effective action, consistent with international law, to deter reflagging of vessels as a means of avoiding compliance with applicable conservation and management rules for fishing activities on the high seas.[51]

46 FAO, 'Strong Support for Firmer Action against IUU Fishing' (Press Release, 3 March 2003) <http://www.fao.org/english/newsroom/news/2003/14660-en.html> at 1 October 2004.

47 It is noted that the practice of reflagging has been in existence since the early 1920s and its attraction to fishing vessel owners has been closely linked to the growth of IUU fishing. The history of re-flagging is discussed in Chapter 2, however a comprehensive analysis is contained in R. Carlisle, *Sovereignty for Sale : The Origins and Evolution of the Panamanian and Liberian Flags of Convenience* (1981). See also B.A. Boczek, *Flags of Convenience- An International Legal Study* (1962)

48 Adopted at the *United Nations Conference on the Environment and Development*, Rio de Janerio, September 1992. Also known as the 'Rio Earth Summit'. A copy of Agenda 21 can be found at <www.un.org/esa/sustdev/documents/agenda21/english/agenda 21chapter17.htm> (May 2004).

49 Ibid Chapter 17, paragraph 17.45.

50 Ibid paragraph 17.52.

51 *Declaration of Cancun*, adopted at the International Conference on Responsible Fishing, Cancun, Mexico, 6-8 May 1992, paragraphs 11 and 13.

oceans.[81] The FAO has estimated that IUU fishing accounts for up to 30% of total marine catches.[82] Preliminary reports by the High Seas Task Force established under the auspices of the OECD suggest that 'the overall value of IUU fish might be somewhere between US$2 billion and US$15 billion per year.'[83]

The *International Commission for the Conservation of Atlantic Tuna* (ICCAT) estimated unreported catches in 1999 accounted for as much as 10% of the overall catches of major Atlantic tuna species.[84] Up to one third of the total allowable catch of southern bluefin tuna for 1999 were thought to have been taken by non members to the *Convention for the Conservation of Southern Bluefin Tuna.*[85] The regions of the Southern Ocean which fall under the *CCAMLR* area of application have similarly been subject to a deluge of IUU fishers. The *CCAMLR* Working Group on Fish Stock Assessment estimated that between 1997-2000, the IUU catch was in the vicinity of 90,000 tonnes, a figure which represents twice the level of the regulated catch.[86]

Even more telling is the fact that the high seas harvest has doubled in the last twenty years of the 20th century, yet domestic fisheries (i.e. coastal State managed fisheries) have recorded a decline. One author has stated this represents a 'spill over' effect of domestic fishing capacity to international high seas fisheries.[87]

It is also important to recognise that the impacts of IUU fishing are not limited to the direct impact on fish stock populations and the livelihood of legitimate fishers. Over the past decade, RFMOs, NGOs and the international community have highlighted the environmental impacts of IUU fishing, including a heightened level of sea bird mortality and the harvesting of juvenile fish. By way of example, it is estimated that in 1988 approximately 50,000 to 89,000 seabirds were caught by IUU fishers. This figure compares unfavourably with the 1,562 seabirds caught by legal fishers.[88] Further environmental damage is caused by the discarded lines and

---

[81] For an overview see, K. Bray, 'A Global Review of Illegal, Unreported and Unregulated (IUU) Fishing' (paper presented at the Expert Consultation on IUU Fishing organised by the Government of Australia in cooperation with the FAO, Sydney, Australia, 15-19 May 2000). <http://www.oceanlaw.net/netpath/page4-iuu.htm>. See also J. Swan, *International Action and Responses by Regional Fishery Bodies or Arrangements to Prevent, Deter and Eliminate Illegal, Unreported and Unregulated Fishing* FAO Fisheries Circular No.996 (2003) part 4.

[82] FAO, *The State of World Fisheries and Aquaculture Report* (2000) 57. FAO Technical Guidelines for Responsible Fisheries (No.9), above, 2, estimates as much as one quarter of the total catch might be attributable to IUU fishing, however also acknowledges that fully reliable data is scarce.

[83] Submission by the High Seas Task Force to the *International Conference on the Governance of High Seas Fisheries and the UN Fish Agreement* (2005) above, 1.

[84] FAO Technical Guidelines for Responsible Fisheries (No. 9), above, Section 1.1.2.

[85] Schmidt, 'Addressing IUU Fishing (2004) above, 4.

[86] Ibid.

[87] Ibid.

[88] 1999 Report of the UNSG on Oceans and law of the sea, above, paragraph 252.

nets of the IUU fishers.[89] The working and living conditions for the crew onboard many IUU vessels have been described as 'bordering on dangerous' by several NGO reports.[90]

Indeed IUU fishing cannot be regarded in simple terms of undesirable fishing activity. There are many factors driving the persistence of IUU fishing, including the economic benefits to be gained by the operators and owners of the vessels. Paying low wages, providing substandard working and living conditions and cutting operating costs by utilising environmentally unfriendly fishing methods are merely components of the economic package.[91]

1.6 The Experience of IUU Fishing in the Southern Ocean

Whilst RFMOs have generally reported high levels of IUU fishing within their respective area of competence,[92] the impact of organised IUU fishing has been felt particularly in the Southern Ocean and within the EEZs located adjacent to a number of sub-Antarctic islands. The remote location of these EEZs, coupled with the freezing and mountainous seas encountered in the Southern Ocean, magnifies the difficulties experienced by coastal State authorities in regulating their maritime zones and deterring IUU fishers. Excepting the pockets of coastal State jurisdiction, the Southern Ocean south of 60 degrees south latitude is governed by the Commission established by the *CCAMLR*.[93] However, States who are not contracting parties to members of *CCAMLR* regard all of the Southern Ocean, with the exception of the remote EEZs, as high seas.[94]

1.6.1 Emerging Concerns about IUU Fishing in *CCAMLR* Waters

Concerns about illegal fishing activities in the Southern Ocean were first raised in 1993. The *CCAMLR* Scientific Committee reported that Toothfish stocks around

89 The lines and nets are often cut away by IUU fishers attempting to outrun coastal state authorities. See also, FAO Technical Guidelines for Responsible Fisheries (No.9), above, Section 1.1, 1.

90 Schmidt, 'Addressing IUU Fishing', above, 3. Schmidt refers to Greenpeace and International Labour Organisation reports of dismal salaries, poor vessel standards and a generally lower level of education amongst employee fishers.

91 See generally, O. Schram Stokke and D.Vidas, 'Regulating IUU Fishing or Combating Operations' OECD, *Workshop on IUU Fishing,* above.

92 For example, see, 1999 Report of the UNSG on Oceans and law of the sea, above, paragraph 250.

93 *CCAMLR*, Article VII.

94 The validity of remote EEZs such as the Heard and McDonald Islands' EEZ has been raised by Judge Vukas in two cases before the International Tribunal for the Law of the Sea ('ITLOS'): see *The Monte Confurco Case (Seychelles v France)* ITLOS Case No 6 (Unreported, Declaration of Judge Vukas, 18 December 2000) (*Monte Confurco Case*); *The Volga Case (The Russian Federation v Australia)* ITLOS Case No 11 (Unreported, Declaration of Vice-President Vukas, 23 December 2002). This point is examined in Part 4.

South Georgia Island may have been depleted to approximately 30 per cent of the original stock levels.[95] Signs of active management appeared on the agenda for the 1995 *CCAMLR* annual meeting.[96] The Working Group on Fish Stock Assessment reported that '[t]he unreported catch was either of the same order or higher than the reported catch'.[97] It was also observed that over the previous four years, the reported catch represented just 40 per cent of the estimated total catch harvested from within sub-area 48.3 (South Georgia Island) and adjacent areas.[98]

During discussions at the 1997 *CCAMLR* meeting, many member states expressed concern for the growing incidence of IUU fishing within the *Convention* Area. The European Community (EC) representative stated that the EC 'considered that [the] *CCAMLR* [Commission] faced a major challenge resulting from the blatantly illegal and/or non-notified fishing activities'.[99] The *CCAMLR* Commission concluded, at the 1997 meeting, that fisheries management is aggravated by the fact that:

> illegal, unregulated fisheries and unreported catches today exceed reported fishing by a factor several times over. No less aggravating is the fact that more than half of the vessels presumed to engage in illegal, unregulated and unreported fishing fly the flags of *CCAMLR* Member States. That underlines the urgent need for *CCAMLR* and *CCAMLR* Member States to bring their own house in order.[100]

1.6.2 IUU Fishing in the *CCAMLR* Waters in the 21st Century

The problem of IUU fishing has continued to perplex the *CCAMLR* contracting parties in the 21st century. The *CCAMLR* Commission noted during the annual Commission meeting in 2003 that 'current levels of IUU fishing are unsustainable'.[101] As a *CCAMLR* contracting party and a State with a declared EEZ in Southern Ocean waters,[102] Australia has significant political and economic interests which are infringed by the persistent practice of IUU fishing. The principal target fish for the IUU fishing industry is the Patagonian Toothfish.

---

95 CCAMLR-XII (1993), Report of the Twelfth Meeting of the Scientific Committee, paragraph 3.34. Note the word 'may' was used as there was in 1993 and remains today, uncertainty about exact population levels. See also Agnew, 'The Illegal and Unregulated Fishery for Toothfish in the Southern Ocean', above n 3, 361, 362. 'Original levels' would appear to mean the first recorded levels.

96 CCAMLR-XIV, Report of the Fourteenth Meeting of the Scientific Committee.

97 Ibid paragraph 4.16.

98 Ibid paragraph 4.17. The 'Fishing Season' referred to in Table 1 runs from 1 December to 30 November.

99 *CCAMLR* Commission, *Report of the Sixteenth Meeting of the Commission*, (1997), paragraph 5.2.

100 Ibid paragraph 5.8.

101 *CCAMLR- XXII,* Report of the Twenty-Second Meeting of the Commission (2003) paragraph 8.3.

102 In fact nearly all of the Heard and McDonald Islands' EEZ is located within the *CCAMLR* area of application. See Map 5.1.

It is difficult to verify estimates of the quantity of Toothfish harvested by the IUU fishing industry. However, IUU fishing as a percentage of the total catch of Toothfish within the *Convention* Area is significant. Table 1.1 below illustrates the extent of the problem raised by IUU fishing. Whilst the peak of IUU fishing appears to have passed, it remains a significant management problem for the *CCAMLR* Commission. The estimates in Table 1.1 for the 2001–02 fishing year indicate a rise followed by a marked decline in 2003–2004.

Fishing Season	1996–97	1997–98	1998–99	1999–00	2000–01	2001–02	2002-03	2003-04
Estimated IUU Catch	32 673	15 106	5868	7644	8802	11 812	10 070	2622
Total Catch	45 130	28 518	19 531	25 214	22 598	27198	26 877	15929
IUU Catch as % of Total	72.4%	53%	30%	30%	39%	48%	37.5%	16.5%

Table 1.1 *CCAMLR* Estimates of IUU Toothfish Fishing (in Tons) within the Convention Area from 1996–97 to 2003–04[103]

1.6.3 The *CCAMLR* Response

In response to this growing threat to the future existence of the Patagonian Toothfish, and indeed, the Convention itself, the *CCAMLR* Commission adopted a number of conservation measures specifically aimed at eliminating or at least deterring IUU fishing within the *Convention* Area. The *CCAMLR* Commission adopted a Catch Documentation Scheme (CDS) in 1999 and implemented an IUU Vessel database (IUU database) in 2002. Both initiatives are aimed at deterring IUU fishing for Patagonian Toothfish within the Convention Area.[104] Whilst the estimates for the 2003-2004 fishing season do indicate a marked decline in the incidence of IUU fishing within the Convention Area, this cannot be attributed wholly to new *CCAMLR* conservation measures.[105] The *CCAMLR* Standing Committee on Implementation and Compliance (SCIC) noted that the Working Group on Fish Stock Assessment had not been able to identify possible reasons for the decline in IUU estimates but considered a number of factors. These were listed as:

103 The information for this table was obtained from *CCAMLR-XXIII* (2004) Report of the Twenty-Third Meeting of the Commission Annex 5, Report of the Standing Committee on Implementation and Compliance (SCIC) paragraph 2.1.

104 *CCAMLR*, Article 1(1). In essence this encompasses the marine living resources of the area south of 60 degrees South latitude including the area between that latitude and the Antarctic Convergence.

105 *CCAMLR- XXII*, (2004) Report of the Standing Committee on Implementation and Compliance, paragraph 2.2.

- The inadequacy of the current level of monitoring, control and surveillance activities in the Convention Area;[106]
- A shifting of IUU fishing activity in the Convention Area to areas outside fishing grounds where licensed vessels operate;
- The depletion of Toothfish stocks;
- Reflagging of vessels to states which do not participate in the CDS, resulting in the receipt of fewer CDS certificates, and
- The impact of conservation measures on the reduction of IUU fishing and the continued monitoring of the world trade in Toothfish.

A detailed review of the *CCAMLR* response to IUU fishing activity within the *CCAMLR* area of application is contained in Part 4.

1.6.4 The Australian Response

The Australian Government has responded to the persistent nature of IUU fishing in its Southern Ocean EEZ by reviewing and amending policy and legislation. Whilst only eight IUU vessels have been arrested to date with the last apprehension occurring in January 2004, these statistics belie the actual numbers of vessels engaging in IUU fishing, not only within the Heard and McDonald Island's fishing zone, but within the *CCAMLR* area of application. In March 2005 the Australian flagged *Oceanic Viking,* a dedicated armed patrol vessel, encountered up to six foreign fishing vessels flagged to *CCAMLR* non-contracting parties.[107] In 1997, before the initial arrests of IUU fishing vessels took place, it was estimated that up to 70 fishing vessels were engaged in IUU fishing in and adjacent to the Heard and MacDonald Island's fishing zone.[108]

In May 2005, the Australian government announced a commitment to continue armed patrols in the Southern Ocean at least until 2010.[109] These patrols complement the policy and legislative review aimed at making IUU fishing an economically unattractive option for IUU fishers and operators. The detailed review of Australian measures to deter IUU fishing is contained in Part 5.

---

[106] Which presumably means IUU activity is going undetected.

[107] I. Macdonald, Minister for Fisheries, Forestry and Conservation and C. Ellison, Minister for Justice and Customs, 'Flag of Convenience vessels flaunt international rules' 4 March 2005. <www.mffc.gov.au/releases/2005>. (4 March 2005).

[108] AFP, 'Fishing Piracy around Antarctica shadows Treaty Meeting' Christchurch, New Zealand, 12 May 1997.

[109] I. Macdonald, Minister for Fisheries, Forestry and Conservation, 'Long-term commitment to Southern Ocean armed patrols' 10 May 2005, <www.mffc.gov.au/releases/2005>. (20 May 2005).

1.7 Why IUU Fishing Developed and Persists

Although this particular question is addressed fully in Part 2, it is appropriate in this introductory Chapter to provide a broad overview of key factors identified as contributing to the development and persistence of IUU fishing. This is because an understanding of the depth of the problem is predicated upon a clear appreciation of these factors, both legal and non-legal.

IUU fishing has been able to persist and arguably, thrive, through the operation of two key principles of international law.[110] The right to the freedom of fishing on the high seas has, combined with the principle of flag state authority allowed fishers to exercise their high seas rights whilst concurrently avoiding regulation if they choose to do so. The FAO has recently noted that 'IUU fishing on the high seas constitutes first and foremost a failure on the part of flag states to abide by responsibilities recognised by the international community and reflected in relevant international instruments'.[111]

The principle of the freedom of the high seas, a subset of which is the freedom of fishing on the high seas,[112] militates against the establishment of an effective regime for high seas fisheries regulation. Whilst the *LOSC* does detail general limitations on the exercise of the right to fish on the high seas, through the application of Articles 116-119, in practice some States do allow their flagged vessels to fish the high seas without ensuring compliance with these general obligations. In the absence of enforceable third party rights, the obligation to ensure high seas fisheries are conducted at the very least in keeping with the intention of the State parties agreeing to the content of Arts 116-119 during *UNCLOS* III,[113] falls upon flag States.[114] In the past decade, the inability or unwillingness of flag States to fulfil their international law obligations, particularly in relation to non-compliant fishing vessel operators and IUU fishers, has become apparent.

Fishers or corporate fishing entities wishing to avoid regulation can conveniently do so through registration with 'suitable' flags. This means registration with States which do not exercise their flag State responsibility over fishing vessels.[115] The term

---

110 Although many more factors have been identified as relevant to the presence and persistence of IUU fishing. These factors are discussed in Part 2 of this thesis.

111 FAO Technical Guidelines on Responsible Fishing (No.9) above, 11.

112 *LOSC*, Article 87(e).

113 The Third United Nations Conference on the Law of the Sea 1973-1982.

114 Flag States are also called upon to ensure their flagged fishing vessels fish responsibly under the FAO *Code of Conduct*. Under the FAO *Compliance Agreement*, a binding agreement which reflects provisions of the *Code of Conduct*, all flag states are obliged to ensure their vessels do not undermine conservation measures. Further, a state is not to grant a fishing authorisation unless it is able to control the fishing operations of a vessel. Under the *UNFSA* the duties of flag states are outlined in Articles 18 and 19.

115 See Schram Stokke and Vidas, 'Regulating IUU Fishing' above, paragraph 22 and generally paragraphs 17-25.

'flag of convenience' has been adopted in the past to label such practices, as has the term 'states with open registries.' This term denotes the lack of a genuine link between the vessel and the flag State. It also indicates that the registration is one of mere convenience to avoid incurring international or regional obligations and/or to take advantage of a particular State's lax attitude with respect to the management of its flagged fishing vessels.[116] The attractiveness of flags of convenience to IUU fishers is self evident. States operating such registries are generally in breach of international obligations, including those arising from the *LOSC*,[117] *UNFSA*,[118] *Compliance Agreement*[119] and *Code of Conduct*[120] and are typically not members of relevant RFMOs.

It has been observed that in 'a world where strong and effective flag state control was exercised, there would be little scope for IUU fishing.'[121] Whilst this is a reasonable observation there is little economic incentive for States to either decline applications for registration or to take effective enforcement action with respect to non-compliant fishing vessels. It involves balancing between the collective interest in sustainable management of high seas fisheries with the individual interests of States. Similarly there are significant economic drivers encouraging IUU fishers to seek registration with 'suitable' States and this fuels the market for flag of convenience States.

It is demonstrated in Part 2 that many of the legal factors are immutable. For example, legal principles of international law underpinning flag state jurisdiction and the freedom of fishing on the high seas have withstood challenges throughout the centuries.[122] Political and economic factors are potentially more malleable and may be addressed through the adoption of measures complying with the international rule of law. The international response to IUU fishing and the framework for legal measures established under international law follows in Part 3.

---

116 See D. Agnew and C. Barnes, 'Economic Aspects and Drivers of IUU Fishing: Building a Framework' (paper prepared for OECD *Workshop on IUU Fishing, Paris*, above) paragraph 19-23. Reproduced in OECD, *Fish Piracy, Combating Illegal, Unreported and Unregulated Fishing* (2004).

117 *LOSC*, Article 94.

118 *UN Fish Stocks Agreement*, paragraphs 18 -19 and 34-50.

119 *Compliance Agreement*, Article 1 contains the general duty to ensure flag vessels do not engage in any activity that undermines the effectiveness of international and conservation measures. Arts 2-8 contain additional responsibilities.

120 *The Code of Conduct*, Articles 6.10 and 6.11. Although the duties outlined in the Code are voluntary, it supports obligations in the Compliance Agreement, *LOSC* and UNFSA.

121 D.J. Doulman, 'A General Overview of some aspects of IUU Fishing' (paper prepared for *OECD Workshop on IUU Fishing*) above.

122 During UNCLOS III, coastal and port states sought to create greater authority over tankers whilst in ports, the territorial sea, EEZ or straits. This was prompted by an increase in tanker accidents and oil spills in the 1960s and 1970s. The flag state group of states argued that universal registration should be achieved via the IMO and withstood the challenge to flag state authority. E.L. Miles, *Global Ocean Politics* The *Decision Making Process at the Third United Nations Conference on the Law of the Sea 1973-1982* (1988) 92-93. High seas freedoms have been preserved in the *LOSC*, Articles 87 and 116.

1.8 Conclusion

The resources of the high seas are threatened by overfishing. In this context IUU fishing is an activity which threatens both the present and future management of sustainable high seas fisheries. The international community has responded to the persistence of IUU fishing by placing the issue on the global agenda and through the adoption of a number of multi-lateral instruments. There are many joint and several factors contributing to the development and persistence of IUU fishing, yet there are options for the international community, RFMOs and coastal States to take action to deter and eliminate the practice in the absence of flag state control whilst also observing both flag state authority and the freedom of fishing on the high seas. Whilst the effects of IUU fishing are global, the Southern Ocean has been targeted specifically by IUU fishers. The response of *CCAMLR* and Australia to adopt measures to deter IUU fishing in their respective maritime jurisdictions in the Southern Ocean form the central focus of this study.

Introduction

A number of interrelated factors have contributed to the emergence, development and persistence of IUU fishing. In order to appreciate the complexity of the problems raised by IUU fishing it is necessary to examine these factors and understand the way in which they have influenced the conduct of marine fisheries generally. The identification of contributing factors has been addressed at the international level, most recently at the 2004 OECD *Workshop on IUU Fishing Activities*, a gathering of experts focused on addressing IUU fishing from a multi-disciplinary approach - in particular by examining the 'economic and social drivers of IUU fishing.'[1] However, this realisation that the task of tackling IUU fishers and operators requires, as a pre-condition, an understanding of the legal and non-legal environment in which fisheries have developed is still gaining wider acceptance. With the exception of conference proceedings, there remains little published work which examines the interrelated factors and draws them together in a comprehensive analysis.[2]

IUU fishing is a complex problem which challenges all persons and organisations concerned in the sustainable management of marine fisheries. There are many factors contributing to its development and persistence. These factors have origins in legal, political, social, economic and even historical interests and often the distinction between interests is blurred. For example the economic and political interests of the Dutch in the freedom of navigation of the high seas and access to the trade in the New World shaped the early Grotian pronouncements on the freedom of the high seas.[3] The principle of the freedom of the high seas and varying expressions of that

1 OECD *Workshop on Illegal, Unreported and Unregulated (IUU) Fishing Activities*, Paris, 19-20 April 2004 <http://www.oecd.org/document/5/0,2340,en_2649_33901_21007109_1_1_1_1,00.html > (12 July 2004). IUU fishing was again the subject of international attention during the *International Conference on the Governance of High Seas Fisheries and the UN Fish Stocks Agreement –Moving from Words to Action*, St John's, Newfoundland and Labrador, Canada, 1-5 May 2005. See Conference home page at <http://www.dfo.gc.ca/fgc-cgp/index_e.htm > (14 May 2005).

2 The author has approached the study of high seas fisheries management from a legal, political and economic perspective for a number of years. See, R. Baird, 'Political and Commercial Interests as Influences in the Development of the Doctrine of the Freedom of the High Seas' (1996) 12 *Queensland University of Technology Law Journal* 274-292 and 'Illegal, Unreported and Unregulated Fishing: An Analysis of the Legal, Economic and Historical Factors relevant to its Development and Persistence' (2004) 5(2) *Melbourne Journal of International Law* 299-335. There are several papers arising from the OECD *Workshop on IUU Fishing* , ibid. Some of these papers have been published by the OECD. See OECD, *Fish Piracy, Combating Illegal, Unreported and Unregulated Fishing* (2004). In particular see the paper by O. Schram Stokke and D. Vidas, 'Regulating IUU Fishing or Combating IUU Operations' which provides an insightful analysis of economic factors.

3 H. Grotius, *The Freedom of the Seas (or the Right which belongs to the Dutch to take part in the East Indian Trade)* (1604) as translated by R.V.D. Magoffin (1916) Introductory Note, pages VI-VII and 38. (Grotius, Freedom of the Seas) Albeit Grotius' early claims were couched in terms of the freedom of navigation and access to trade.

freedom are now accepted principles of customary law[4] and have been described as 'representing a sensible and wholesome concept of shared used.'[5]

However, this shared use, with regard to fisheries, requires the maintenance, at the very least, of a semblance of order so that what is free for all to take is taken with regard to present and future generations.[6] Therein lies the problem. In the absence of a regulatory authority to govern the high seas, responsibility has fallen upon the flag State to exercise control over its fishing vessels.[7] The experience, particularly since the conclusion of UNCLOS III, illustrates that more often than not, flag States have been either unwilling or unable to effectively regulate the activities of their fishing vessels on the high seas.[8] This lack of control has facilitated an increase in IUU fishing. Global marine catches have trebled since 1950 and the percentage of fish stocks estimated to be in an overexploited state continues to rise.[9] This in turn creates an environment of increased competition for marine fish stocks in which more, not less flag State control is required. Within this environment the phenomenon of IUU fishing has emerged as a matter of serious and increasing concern.[10]

Concealed within this brief précis of the development of IUU fishing is the myriad of underlying legal and non-legal factors that have contributed to its development. The displacement of distance water fishing vessels through the extension, by many States, of coastal maritime zones; rapid advancements in technology, which have improved vessel efficiency; significant capital investment in the fishing industry (including subsidies); increases in the global population and corresponding increases

---

4 See specifically Article 87 of the *United Nations Convention for the Law of the Sea* opened for signature 10 December 1982, 19 ILM 1261 (entered into force 16 November 1994) ('*LOSC*'). See also Arts 86-121. The *LOSC* has been ratified or acceded to by 148 states (as of 11 March 2005). The concept was articulated in Art 2 of the Convention on the High Seas 1958 and applied by the International Court of Justice (ICJ) in the *Anglo-Norwegian Fisheries Case* ICJ Reports (1951) 116.

5 I. Brownlie, *Principles of Public International Law* (6th ed 2003) 225.

6 The concept of intergenerational equity is arguably emerging as a matter of customary international law. It has been expressed in Principles 1-2 of the Stockholm Declaration of the United Nations Conference on the Human Environment (1972) and Principle 3 of the Rio Declaration on Environment and Development (1992). The ICJ also supported the concept in its Advisory Opinion in the *Legality of the Threat or Use of Nuclear Weapons* (1996) ICJ Rep. 241-242. The concept is linked to the principle of sustainable development which has wide international support. See the International Law Association's New Delhi Declaration of Principles of International Law relating to Sustainable Development (2002).

7 It is noted that historically vessels sailed under the flag of a State in order to enjoy the benefits of protection of that State during commerce and exploration. See D. Caron, 'Ships, Nationality and Status' *Encyclopaedia of Public International Law* Vol. 4 (2000), 400.

8 The duties of flag States and flags of convenience are examined further below in this Chapter.

9 See Chapter 1 and Graph 1.1.

10 *International Plan of Action to Prevent, Deter and Eliminate Illegal, Unreported and Unregulated Fishing* (*IPOA-IUU*) adopted by the 25th Session of the FAO Committee on Fisheries (COFI) on 2 March 2001, paragraph 1.

in human consumption of fish; the low risk of apprehension for those considering undertaking illegal fishing; and an over capacity in the global fishing fleet must be considered. Marine fisheries in the 21st century have become characterised by too many vessels — and efficient ones at that — pursuing fewer fish in a greatly reduced area.[11] When the cumulative impact of these political, historical and economic and social factors is assessed, one appreciates the profound way in which the nature of fishing, particularly in the latter half of the 20th century, has been altered. This shift is comparable to the transformation of marine fisheries from a 'trade to a science'.[12]

Whilst it is impossible to retreat from the 21st century practice of marine fisheries, one can learn from its evolution. What follows is an analysis of the development of marine fisheries in the 20th century and the factors contributing to the emergence of IUU fishing. International, regional and national responses and regional and coastal State efforts to deter IUU fishing are examined in Parts 3, 4 and 5 respectively.

11 See J. Diouf, FAO Director General, 'Too Many Vessels Chasing Too Few Fish' (Press Release, 1 October 2001) <http://www.fao.org/WAICENT/OIS/PRESS_NE/PRESSENG/2001/pren0158.htm> (1 October 2004). Diouf stated: 'Consequently today there are too many vessels chasing too few fish'.

12 D. Hunter, J. Salzman and D. Zaelke, *International Environmental Law and Policy* (2nd ed, 2002) 674.

Chapter 2

Legal Factors Contributing to the Development of IUU Fishing

2.1 Legal Factors influencing the Development and Practice of High Seas Fisheries and the Emergence of IUU Fishing

There are a number of legal principles which shape the conduct of high seas fisheries. These legal principles are fundamental pillars of the international law of the sea and are discussed individually below. One observation uniformly applicable to these legal factors is that as principles of customary international law their development and acceptance has been an incremental process. Change to customary international law is a slow process and indeed only a remote possibility.[13] Currently IUU operators can, and do, exploit limitations in the international law of the sea to avoid the application of regional and national fishing measures. Furthermore, the practice of IUU fishing is fluid and can be quickly adapted to take advantage of changing circumstances and avoid legal or policy measures implemented by national governments.[14]

This Chapter reviews the legal principles identified as relevant to the development of IUU fishing and concludes that to effectively deter IUU fishing, these legal limitations need to be acknowledged and in the absence of any amendment to international law,[15] alternative strategies or mechanisms to deter and possibly eliminate the practice of IUU fishing, need to be implemented. It may be trite to label these fundamental legal principles as factors contributing to the development of IUU fishing however it is argued that whilst they are undoubtedly important rules of law which have, and continue to contribute to the stability of international law of the sea, their operation also enables 'free riders' to avoid responsibility. Those seeking to identify effective measures to deter IUU fishing need to first appreciate the legal environment in which it operates. For example, the rule of *pacta tertii* is a

[13] Discrete discussions on the freedom of fishing and flag State authority below illustrate the degree to which these principles are entrenched in customary international law.

[14] This view is supported by Stokke and Vidas, 'Regulating IUU Fishing or Combating IUU Operations' above, paragraph 51.

[15] Whilst it is outside the scope of this text to discuss the likelihood of amendment to international law principles, it is the author's view that amendment (as distinct from limitation or modification) to fundamental principles such as flag State authority, freedom of the high seas and *pacta tertii* is not likely in the foreseeable future.

central tenant of customary international law and this rule has been used in practice by third States to fish the high seas without incurring any of the treaty obligations specifically aimed at facilitating responsible high seas fishing.

2.2 The Development of the Freedom of the High Seas

2.2.1 The Influence of Hugo Grotius

The expression, 'freedom of the high seas' is inextricably linked with the writings of Hugo Grotius.[16] However, there is evidence of dialogue on the legal status of the seas during the ancient Greek and Roman Empires.[17] Long before Grotius penned his impassioned defence of the freedom of the seas, disputes over access to traditional fishing grounds in coastal waters were common place. As fishing vessels were limited in range and techniques were primitive, high seas fisheries were virtually non existent. The early fisheries disputes were therefore over coastal fisheries and based upon economic factors - usually the access of local fishers to coastal fisheries. Disputes of this nature continued after the 1609 publication of Grotius' *Mare Liberum*.

Attempts to limit the access by foreign fishers to waters adjacent to the British coastline are apparent from as early as 1360. English Kings Edward III and Henry V, VI and VII made various proclamations asserting the rights of British fishers to coastal waters and seeking to exclude foreign fishers, the bulk of whom were Dutch.[18] However, foreigners were from time to time allowed limited access. Licences were issued to fishers from the 'low countries' to harvest herring from British waters.[19]

Early declarations of the freedom of the seas in the late 1400's were directed at securing access to maritime trade routes that were, at the time, being monopolised by the major sea powers, Spain and Portugal.[20] In the 1600's unrestricted access to the

---

16 Grotius, *The Freedom of the High Seas*, above.

17 The Roman Jurist Marcianus is credited with the first recorded written pronouncement on the legal status of the seas. He declared the seas and the fish within them were *communis omnium naturali jure* or 'common or open to all men by operation of natural law.' P.T. Fenn, 'Justinian and the Freedom of the Sea' (1925) 19 *American Journal of International Law* 717, 718 and 723-726. In reality the supremacy of Rome meant the Mediterranean was regarded as a Roman lake- the free use of which was open to all Roman citizens.

18 J.R. Elder, *The Royal Fisheries Companies of the 17th Century* (1912) 4.

19 Ibid. 21 year licences were issued to fishermen from Holland, also referred to as the 'low countries.'

20 The Papal Bull of 1493 endorsed the naval power of Spain and Portugal and their monopoly over vast newly discovered tracts of the Atlantic and Indian Oceans. The Treaty of Tordesillas of 7 June 1494 confirmed the division of the New World between the two States. Through military power and with religious support, Spain and Portugal sought to exert private property rights over the Atlantic and Indian Oceans and to deny access to other States. See also, G. Knight, *Managing the Sea's Living Resources* (1997) 14-15.

oceans guaranteed rights of navigation, access to lucrative trade routes in the Americas, and to a lesser extent, access to fisheries. The early attempts to control access by foreign fishers to coastal waters can be viewed therefore as a local issue that became subsumed by the wider national interests of the British and Dutch in challenging the Spanish and Portuguese monopoly over trade and navigation to the 'New World.'

Over the ensuing four centuries, the right to fish the high seas has become entrenched in international practice and law as a principle of customary international law. The analysis supports the conclusion that fisheries law evolved in the wake of the settlement of the freedom of navigation on the high seas. During this same period, and perhaps as a reflection of the political climate of nation building and colonisation, coastal State jurisdiction over an adjacent sea corridor was also internationally recognised and later enshrined within the *LOSC*.

Shortly after the formation of the Dutch East India Company, Grotius was requested to write a treatise on the law of prize. The request arose out of the need to defend the capture by a Dutch East India Company vessel of a Portuguese galleon in the straits of Malacca in 1602.[21] *De Jure Praedae* was completed in 1605 but not immediately published.

In November 1608, Chapter XII was published anonymously.[22] Entitled *Mare Liberum* or 'The Freedom of the Seas or the Right which belongs to the Dutch to take part in the East Indian Trade', the Chapter was written specifically to refute claims by Portugal and Spain to the control of the high seas.[23] Political and commercial interests dictated the decision to publish the Chapter in isolation. It is probable that if the entire treatise had been published as written in 1608, the impact of *Mare Liberum* may have been lost within the wider arguments on the law of prize contained therein.

Grotius drew a distinction between the limitless nature of the sea and the finite nature of land. The sea was common to all and open to use by all. Grotius stated:

> [t]hat which cannot be occupied, or which never has been occupied, cannot be the property of anyone, because all property has arisen from occupation.[24]

A distinction was also made between things which were *res nullius* (belonging to no one) and *res communis* (a common possession). Objects which belong to no one yet are capable of being seized and possessed, are distinguished from objects which have been marked out for common use (fish, wild animals and birds).[25] The sea by its nature not only "allows but enjoins"[26] common use. Hugo Grotius further

---

21 Grotius, *The Freedom of the Sea*, above, Introductory Note, VI-VII.

22 Ibid, V. Though it is arguable that Grotius was requested to publish this Chapter on the freedom of the seas to refute the continued claims of Spain and Portugal.

23 Ibid, VIII.

24 Ibid, 27.

25 Ibid, 29.

26 Ibid, 30.

stressed that the sea being 'common to all' could not in any part be considered the territory of any people. He contrasted this with public property which a nation could occupy for its citizens to ensure common use.[27] The Roman maritime supremacy and use of fleets to protect navigation and capture pirates was explained as the exercise of a common right to use the seas, not as evidence of the existence of a private right.[28]

Critical of the claims of the Portuguese and Spanish to exclusive trade rights and access to the expanse of sea between the old and new worlds, Grotius wrote:

> If in a thing so vast as the sea, a man were to reserve to himself from general use nothing more than mere sovereignty, still he would be considered a seeker after unreasonable power. If a man were to enjoin other people from fishing, he would not escape the reproach of monstrous greed. But the man who even prevents navigation, a thing which means no loss to himself, what are we to say of him.[29]

A direct reference to the supposed inexhaustible nature of marine resources was made within *Mare Liberum.* Grotius reiterated that the fish in a river might be appropriated as they were easily exhaustible. However, he thought such a proposition was impossible in the case of the sea.[30] He did acknowledge that fish within the seas could be appropriated as private property if caught by persons exercising a right of common use.[31]

The reaction to the content of *Mare Liberum*, in particular by Britain, and the long term effects on international law, make *Mare Liberum* the focus of considerable discussion. Yet when considered in its historical and political context, Hugo Groitius' publication can also be appreciated as a development in Dutch foreign policy. In 1581, Holland gained her independence from Spain and sought to engage in the sea trade in her own right. Just 27 years later *Mare Liberum* was published. The British perception of the 'low countries' as a poor neighbour and a dependency of Spain quickly changed. As a newly developing maritime power on Britain's doorstep, Holland represented a threat to British maritime supremacy in the region.[32]

Whilst Holland was working towards independence from Spain and asserting her rights to navigate the seas freely, Britain had enjoyed two centuries of freedom in relation to the use of and access to the seas - at least the seas surrounding the British Isles.[33] The evidence suggests this freedom was asserted because it coincided with

---

27 Ibid, 34.

28 Ibid, 35.

29 Ibid, 38.

30 Ibid, 58. This point was made earlier by Spanish jurist Ferdinand Manchea Vasquez.

31 Ibid, 30. This approach was followed in early British and American case law and remains good law. See, eg, *Young v Hichens* (1844) 6 QB 606. *State of Ohio v Shaw* (1902) 65 NE 875.

32 Elder, *The Royal Fisheries Companies*, 5-6.

33 For a discussion see G. Smith, 'The Politics of Law Making: Problems of International Maritime Regulation, Innocent Passage v Free Transit' (1976) 37 *University of Pittsburgh Law Review* 491.

British political motives to ensure freedom of trade and fishing, rather than arising out of idealistic motives to develop the law of the sea.[34] The freedom was extended at times to foreigners, with the Tudor reign noted in history as being synonymous with freedom of the seas.[35]

As Spain and Portugal asserted trade monopolies over the sea, the freedom to fish, navigate and engage in trade became very important elements of British foreign policy. However, Britain also became increasingly concerned by the presence of Dutch fishermen in the North Sea and the arguments made by Grotius on behalf of the Dutch East India Company. Hence British resolve to protect what were regarded as England's maritime resources in the North Sea, firmed. This coincided with the accession of James I, whose Scottish heritage influenced him to favour a more restrictive approach towards foreign fishers.[36] In May 1609, James I proclaimed in essence that:

> [No] person of what nation or quality soever, being not our natural born subject, be permitted to fish upon any of our coasts and seas,... until they have demanded and obtained licences from us.[37]

In response, the Dutch fishers refused to pay any duties to catch herring in the North Sea and continued to maintain a presence in the Sea.[38] In 1635, Charles I sought to reassert British authority over adjacent seas, in particular Scottish seas, by demanding payment from foreign fishers.[39]

In 1613, just five years after the publication of *Mare Liberum,* objection to the Dutch presence in the North Sea compelled the commissioning of two works. *The Abridgement of all the Sea-Laws* by William Wellwood was published in 1613 and *Mare Clausum* by John Selden followed in 1618.[40] The latter work was not published until 1635, by which time the Dutch were taking herring right up to the low water mark off the British coastline.[41]

---

[34] Ibid, 492.

[35] T. W. Fulton, *The Sovereignty of the Seas* (first published 1911, reprinted 1976) 118. Queen Elizabeth (1533-1603) is reported to have refrained from claiming sovereignty of the sea although she asserted the freedom of the seas for both fishing and navigation for British fishermen against the policies of Denmark and Spain.

[36] James I of England was James VI of Scotland. On succeeding Elizabeth I to the throne of England, he brought with him Scottish laws and traditions. One such law was the notion of *landkenning* which meant the King of Scotland possessed whatever lay within the range of vision of a ship itself within sight of the Scottish coastline. Scottish fishermen were more reliant upon adjacent herring fisheries for domestic consumption than the English and the export revenue was valued by the Scottish crown. Further, Scotland's coastline meant the herring fishery was located closer to land in lochs and firths and the intrusion of foreign fishermen was resented. D.P. O'Connell, *The International Law of the Sea* Vol. II (edited by I. Shearer) (1984) Vol.1, 3.

[37] Fulton, above, 150. Though a general prohibition, it was understood to be aimed at Dutch fishers.

[38] Ibid.

[39] Ibid.

[40] H. Vreeland, *Hugo Grotius-The Father of the Modern Science of International Law* (first published 1917, reprinted in 1986) 58.

[41] Fulton, above, 150.

The reference to a limitation upon the exercise of high seas freedoms in Article 2 of the *High Seas Convention* is significant. With respect to the freedom of fishing, whilst the exercise of that freedom has become more restricted through both a reduction in the area of high seas via the extension of coastal State jurisdiction over marine resources in the EEZs[51] and the adoption of Articles governing the conservation and management of the living resources of the high seas,[52] it is unlikely that the freedom of fishing in the high seas will be surrendered absolutely. That said, the problems raised by IUU fishing require a regional approach (complementing the international and national responses) and it is the author's view that regional fisheries management should be pursued vigorously. The several measures employed by *CCAMLR* and Australia are supportive of such a regional approach. Furthermore, if responsible fishing upon the high seas cannot be ensured through flag State control, then it is suggested that consideration may be given to further modifying the freedom of fishing on the high seas so that it is exercisable only through membership of a relevant RFMO.[53]

2.2.3 Achieving responsible fishing on the high seas

The freedom of high seas fishing and the authority of flag State jurisdiction over flagged vessels on the high seas are both fundamental principles of international law of the sea and factors facilitating the development of IUU fishing. Responsible fishing on the high seas has not been realised under these principles, for some fishing vessel operators have exploited both the freedom to fish and the lack of flag State action to enforce even minimum international obligations. It would be simplistic to assert that these two principles alone contribute to the development of IUU fishing and this Part examines, in the following pages, the many factors, legal and otherwise, which have contributed to an environment which is conducive to IUU fishing.

There are general limitations on the freedom of fishing the high seas contained in Articles 117-119 of the *LOSC*. However it is argued that the obligations cast in these Articles are of such a general nature that in practice there has been little impact on flag States that are not already inclined to implement the concept of responsible fishing on the high seas throughout their flagged fishing fleet.[54] That

---

51 *LOSC,* Part V.

52 *LOSC,* Articles 116-119.

53 In essence the *UN Fish Stocks Agreement* seeks to establish such a regime where by States which seek to fish within a regional fishery yet refuse to join or cooperate with the relevant RFMO or arrangement are excluded from a quota allocation. However this regime remains based upon consent.

54 This view is supported by Molenaar in the context of the obligation to cooperate on the high seas and/or for transboundary stocks. He states the form of such cooperation is at their discretion and moreover that the consequences for a failure to cooperate are not clearly spelled out in the *LOSC*. E. J. Molenaar, 'Regional Fisheries Management Organisations: Issues of Participation, Allocation and Unregulated Fishing' in A.G. Oude Elferink and D.R. Rothwell (eds) *Oceans Management in the 21st Century: Institutional Frameworks and Responses* (2004) 69-86, 82.

is not to say that the obligations cast in the *LOSC*, such as the duty to cooperate with other States under Articles 117 and 118, do not have the force of customary international law.[55] What is argued is that notwithstanding the customary and treaty law obligation to cooperate in the conservation and management of living resources of the high seas, there is a gap between cooperation and actual results, a gap which can be exploited.

Cooperation may come in many forms[56] and the general nature of the obligation to cooperate in the *LOSC*, without the added obligation to reach agreement on measures for the conservation and management of living resources, means that it is a difficult task to argue, let alone establish, that a State is not cooperating. It is also argued that whilst the *UN Fish Stocks Agreement*, which is reviewed in Chapter 4, does seek to compel States to cooperate, principally via Articles 8 and 17; the influence of this Agreement remains limited. As of May 2005 there were 52 member States to the *UN Fish Stocks Agreement*.[57] Furthermore, it has been ratified by just two of the top ten producing States for marine and inland capture fisheries in 2002 (as identified by the FAO).[58]

Thus, whilst it can be argued that the duty to cooperate in the management of high seas fisheries is a principle of customary international law and moreover the duty to pursue cooperation through RFMOs has also become a matter of custom,[59] compelling States first, to come to the negotiating table and second, to reach consensus, has proven problematic. This is evident in the Southern Blue Fin Tuna dispute between Australia and New Zealand, and Japan which is examined in Part 3 where the duty to cooperate (in that context in respect of an unresolved dispute over tuna catch quotas) was considered by ITLOS. Very recently the UNGA confirmed the duty to cooperate in the conservation and management of straddling fish stocks and highly migratory fish stocks particularly through involvement in RFMOs.[60]

In addition to endorsing the principle of the freedom of the high seas, the ILC considered the question of responsibility for regulating the high seas. The following sections review the legal basis for flag State responsibility, the issue of the grant of nationality to ships and whether the grant is open to third State scrutiny and finally, the way in which flags of convenience undermine the goal of responsible fishing on the high seas.

---

55 This point is discussed further in Chapter 4.

56 As acknowledged by Burke, who notes that it is 'not exhausted by any single one activity'. W.T. Burke, *The New International Law of Fisheries: UNCLOS 1982 and Beyond* (1994) 122.

57 See Annex A, Chapter 4.

58 See Table 4.2, Chapter 4.

59 This argument is substantiated in Chapter 4.

60 UNGA Resolution 59/25 17 January 2005, Preamble and paragraphs 50-57.

has allowed unregulated fishing to continue unchecked for centuries. The resultant environment has been described in the past as the tragedy of the commons.[72]

The unregulated arrangement works well until such time as the demands of humankind and the capacity of the environment to yield to those demands come into equilibrium. With no entity in existence with the mandate to regulate to maintain that equilibrium, the environment suffers and each individual share or return decreases as the environments capacity to produce also decreases.[73] This model depicts the story of high seas fisheries, arguably the one significant unmanaged commons in the world.

The Southern Ocean can be described as a textbook example of the consequences of an unmanaged commons.[74] The high seas fishing grounds for the Patagonian Toothfish are remote, the seas are hostile and the weather conditions are extreme. The financial returns are high[75] the risk of apprehension is low and access is simply a matter of setting a course for the Southern Ocean. In the absence of membership to *CCAMLR* the relevant RFMO, the only authority which might be exercised over fishing vessels on the high seas is by the flag State.

The question is then raised as to whether there is any means by which flag States might be brought to account for failure to adequately regulate their fishing vessels. History has shown that the international community has been unsuccessful in several attempts to create normative requirements, in the context of a genuine link, between vessel and flag State, for the exercise of flag State authority.[76] However, the broader view of the concept of a genuine link, as captured by Meijers in his text *The Nationality of Ships*, would suggest that the requirement of the necessary authority of flag States is fundamental to the maintenance of public order at sea and that the

---

[72] This term was first coined in 1968 by biologist G. Hardin. Hardin stated: 'Ruin is the destination towards which all men rush, each pursuing his own interests in a society that believes in the freedom of the commons. Freedom in a commons brings ruin to all.' G. Hardin, 'The Tragedy of the Commons' (1968) 162 *Science* 1234, 1244.

[73] Hardin, Ibid and see C. Glover, *The End of the Line- How overfishing is changing the world and what we eat* (2004) 133-134. It is noted that Hardin did observe that a commons could be managed successfully by limiting both the number of commoners and access and instituting penalties to act as a deterrent. High seas fisheries did not possess these characteristics.

[74] Glover, ibid, 135.

[75] A single voyage can return over one million dollars worth of Toothfish. It has been observed that 'perhaps the easiest way to make a million bucks at the moment is to put together a boat and go fishing for a season in the Southern Ocean.' See, ABC, 'The Toothfish Pirates' *4 Corners*, 30 September 2002. <www.abc.net.au/4corners/archives/2002b_Monday30September2002.htm> (14 October 2002).

[76] As mentioned in the preceding paragraph, attempts to create both soft law and hard law requirements have not adequately addressed the issue of flag State responsibility. International instruments seeking to improve flag State compliance and the fulfilment of flag State duties are examined in Chapter 4. However, as stated, this study is premised on the basis that immediate action by RFMOs and coastal States is required in the absence of flag State enforcement in preference to devoting more effort and time to persuading recalcitrant flag States to take measures to comply with international and regional fishing measures.

is not to say that the obligations cast in the *LOSC*, such as the duty to cooperate with other States under Articles 117 and 118, do not have the force of customary international law.[55] What is argued is that notwithstanding the customary and treaty law obligation to cooperate in the conservation and management of living resources of the high seas, there is a gap between cooperation and actual results, a gap which can be exploited.

Cooperation may come in many forms[56] and the general nature of the obligation to cooperate in the *LOSC*, without the added obligation to reach agreement on measures for the conservation and management of living resources, means that it is a difficult task to argue, let alone establish, that a State is not cooperating. It is also argued that whilst the *UN Fish Stocks Agreement*, which is reviewed in Chapter 4, does seek to compel States to cooperate, principally via Articles 8 and 17; the influence of this Agreement remains limited. As of May 2005 there were 52 member States to the *UN Fish Stocks Agreement*.[57] Furthermore, it has been ratified by just two of the top ten producing States for marine and inland capture fisheries in 2002 (as identified by the FAO).[58]

Thus, whilst it can be argued that the duty to cooperate in the management of high seas fisheries is a principle of customary international law and moreover the duty to pursue cooperation through RFMOs has also become a matter of custom,[59] compelling States first, to come to the negotiating table and second, to reach consensus, has proven problematic. This is evident in the Southern Blue Fin Tuna dispute between Australia and New Zealand, and Japan which is examined in Part 3 where the duty to cooperate (in that context in respect of an unresolved dispute over tuna catch quotas) was considered by ITLOS. Very recently the UNGA confirmed the duty to cooperate in the conservation and management of straddling fish stocks and highly migratory fish stocks particularly through involvement in RFMOs.[60]

In addition to endorsing the principle of the freedom of the high seas, the ILC considered the question of responsibility for regulating the high seas. The following sections review the legal basis for flag State responsibility, the issue of the grant of nationality to ships and whether the grant is open to third State scrutiny and finally, the way in which flags of convenience undermine the goal of responsible fishing on the high seas.

---

55 This point is discussed further in Chapter 4.

56 As acknowledged by Burke, who notes that it is 'not exhausted by any single one activity'. W.T. Burke, *The New International Law of Fisheries: UNCLOS 1982 and Beyond* (1994) 122.

57 See Annex A, Chapter 4.

58 See Table 4.2, Chapter 4.

59 This argument is substantiated in Chapter 4.

60 UNGA Resolution 59/25 17 January 2005, Preamble and paragraphs 50-57.

2.2.4 Regulating the Freedom of the High Seas

Flag State Responsibilities

The ILC observed in its 1955 Report to the UN General Assembly covering the work of its seventh session that, 'any freedom that is to be exercised in the interests of all entitled to enjoy it must be regulated.'[61] This statement was made in relation to the freedom of the high seas, in the context of positive international laws which acted to safeguard the exercise of the freedom. The ILC listed, in its 1955 Report, five broad issues which were applicable to the exercise of high seas freedoms. These points were not elaborated upon in either Draft Article 27[62] or Article 2 of the *High Seas Convention*.[63] The first issue identified by the ILC to be included as a rule of positive international law was the right of States to exercise their sovereignty on board ships flying their flag. The rationale for flag State sovereignty is that 'the absence of any authority over ships sailing the high seas would lead to chaos.'[64]

The need for order on the high seas was acknowledged by Colombos in 1967 when he stated, 'There need to be some rules governing the acts of those who go down to the sea in ships and occupy their business in great waters.'[65] Further, it has been observed that '[t]he denial to any particular State of authority over the high seas made the authority exercised by flag States over its vessels the linchpin of public order on the high seas.'[66]

The *LOSC* does, under Part VII, Section 2, contain a number of Articles which provide limited guidance with respect to the conservation and management of the living resources of the high seas. However, these Articles refer to cooperation amongst States whose nationals (as distinct from flagged vessels)[67] exploit identical living resources or different living resources in the same area[68] with the aim of negotiating necessary measures for the conservation of high seas living resources

61 ILC Report to the United Nations General Assembly 1955, Vol II. Yearbook of the International Law Commission 19, 22. (ILC 1955 Report to the UNGA)

62 Ibid.

63 Ibid.

64 Ibid.

65 C. J. Colombos. *The International Law of the Sea* (6th ed 1967) 62.

66 Caron, 'Ships, Nationality and Status' above, 400. See also, M.S. McDougal and W.T. Burke, *The Public Order of the Oceans: A Contemporary International Law of the Sea* (1962) 794.

67 This point is made because whereas previously there often was a link between the flag of a ship and the nationality of the crew, the converse is true today. Thus States must not only ensure their flag vessels (often crewed by non nationals) but also their nationals, who may be dispersed far across the global on many vessels (as is the case with Spanish and Russian nationals), engage in fishing on the high seas in a responsible manner. Flag state responsibility is dealt with in seperate Articles and therefore arguably is not a limitation upon the freedom of the high seas but an acknowledgement that flag state authority is necessary for public order on the high seas.

68 *LOSC*, Articles 117-118.

and establishing subregional or regional fisheries organisations. Whilst States have cooperated to establish more than 50 regional fishery bodies,[69] the achievement of effective flag State responsibility over fishing vessels is a separate issue and the efforts of RFMOs are often undermined by flag of convenience States.

Flag State duties are included within a separate grouping of Articles under Section 1 of Part VII, of the *LOSC*. Article 94 of the *LOSC* obligates flag States to effectively exercise jurisdiction and control in technical and social matters over ships flying their flag. With the exception of the general duty to 'assume jurisdiction under its internal law over each ship flying its flag', the specific duties articulated in Art 94(2) and (3) are not directly linked to the control of activities of fishing vessels. These Articles address matters such as the seaworthiness of ships, navigation and safety and the observance of international obligations concerning safety of life at sea, the prevention of collisions, the prevention, reduction and control of marine pollution and the maintenance of communications by radio. Subsequent international instruments have sought to clarify the nature of flag State responsibility, in relation to fishing vessels, through a layered approach of both soft and hard law.[70] These measures are examined in Chapter 4.

The Tragedy of the Commons

The combination of the acknowledged priority of flag State authority and the lack of flag State control[71] in relation to regulating the management of, and access to, high seas fish stocks (including highly migratory species and straddling fish stocks)

---

69 FAO, Regional Fishery Bodies, <http://www.fao.org/fi/body/rfb/chooserfb.htm> at 23 April 2005. The recent comments of Molenaar that not all RFO qualify as RFMOs are relevant as well. Some RFO such as scientific advisory bodies cannot impose legally binding measures (in relation to the conservation and management of marine living resources) on members. See E. Molenaar, 'New areas and Gaps- How to Address them' May 2005 (paper delivered at a *Conference on The Governance of High Seas Fisheries and the UN Fish Stocks Agreement*) above).

70 See the *Agreement to Promote Compliance with International Conservation and Management Measures by Fishing Vessels on the High Seas* 1994 ('*the Compliance Agreement*') which entered into force on 24 April 2003 and is binding upon member States. *The Code of Conduct for Responsible Fisheries* adopted by the 28th session of the FAO Conference 31 October 1995 ('*Code of Conduct* ') and the *IPOA-IUU* are examples of soft law obligations.

71 It is noted that the Australian Minister for Fisheries, Forestry and Conservation has mooted the possibility of taking legal action against flag of convenience States. See, I. Macdonald, Minister for Fisheries, Forestry and Conservation, 'Australia leads fight against Flags of Convenience fishing' 4 May 2005. <www.mffc.gov.au/releases/2005/05073m.html> (5 May 2005). Further, the UNGA has invited the IMO to consider the potential consequence of non-compliance with duties and obligations of flag States in relevant international instruments. UNGA Resolution 59/25, 17 January 2005, paragraph 30. Whilst it is argued that the duty to cooperate in the management of high seas living resources is a principle of customary international law, it is not within the scope of this thesis to explore ways in which flag States might be brought to account for failing to discharge that duty.

has allowed unregulated fishing to continue unchecked for centuries. The resultant environment has been described in the past as the tragedy of the commons.[72]

The unregulated arrangement works well until such time as the demands of humankind and the capacity of the environment to yield to those demands come into equilibrium. With no entity in existence with the mandate to regulate to maintain that equilibrium, the environment suffers and each individual share or return decreases as the environments capacity to produce also decreases.[73] This model depicts the story of high seas fisheries, arguably the one significant unmanaged commons in the world.

The Southern Ocean can be described as a textbook example of the consequences of an unmanaged commons.[74] The high seas fishing grounds for the Patagonian Toothfish are remote, the seas are hostile and the weather conditions are extreme. The financial returns are high[75] the risk of apprehension is low and access is simply a matter of setting a course for the Southern Ocean. In the absence of membership to *CCAMLR* the relevant RFMO, the only authority which might be exercised over fishing vessels on the high seas is by the flag State.

The question is then raised as to whether there is any means by which flag States might be brought to account for failure to adequately regulate their fishing vessels. History has shown that the international community has been unsuccessful in several attempts to create normative requirements, in the context of a genuine link, between vessel and flag State, for the exercise of flag State authority.[76] However, the broader view of the concept of a genuine link, as captured by Meijers in his text *The Nationality of Ships*, would suggest that the requirement of the necessary authority of flag States is fundamental to the maintenance of public order at sea and that the

---

72 This term was first coined in 1968 by biologist G. Hardin. Hardin stated: 'Ruin is the destination towards which all men rush, each pursuing his own interests in a society that believes in the freedom of the commons. Freedom in a commons brings ruin to all.' G. Hardin, 'The Tragedy of the Commons' (1968) 162 *Science* 1234, 1244.

73 Hardin, Ibid and see C. Glover, *The End of the Line- How overfishing is changing the world and what we eat* (2004) 133-134. It is noted that Hardin did observe that a commons could be managed successfully by limiting both the number of commoners and access and instituting penalties to act as a deterrent. High seas fisheries did not possess these characteristics.

74 Glover, ibid, 135.

75 A single voyage can return over one million dollars worth of Toothfish. It has been observed that 'perhaps the easiest way to make a million bucks at the moment is to put together a boat and go fishing for a season in the Southern Ocean.' See, ABC, 'The Toothfish Pirates' *4 Corners*, 30 September 2002. <www.abc.net.au/4corners/archives/2002b_Monday30September2002.htm> (14 October 2002).

76 As mentioned in the preceding paragraph, attempts to create both soft law and hard law requirements have not adequately addressed the issue of flag State responsibility. International instruments seeking to improve flag State compliance and the fulfilment of flag State duties are examined in Chapter 4. However, as stated, this study is premised on the basis that immediate action by RFMOs and coastal States is required in the absence of flag State enforcement in preference to devoting more effort and time to persuading recalcitrant flag States to take measures to comply with international and regional fishing measures.

genuine link concept exists for the sole purpose of safeguarding that authority.[77] The following section reviews the historical basis of the nationality of ships and flag State authority and concludes that the genuine link concept must stand for more than simply safeguarding the authority of the flag State. It is also concluded that the debate as to what constitutes a genuine link is complex and unlikely to be resolved in the foreseeable future.[78] This fact further underscores the importance of identifying alternative measures to deter IUU fishing.

2.2.5 Flag State Authority

A corollary of the freedom of the high seas and the right of all States to grant nationality to a vessel is that the flag State has exclusive jurisdiction over the vessel on the high seas to the extent permissible at law. Flag State authority has, since 1956, been closely linked to the term 'genuine link' although attempts since that date to elucidate precisely what this term entails have had limited success. Whilst this section provides an historical account of the principles governing the grant of nationality to vessels and the adoption of the genuine link concept by the ILC, it does not seek to, nor rest upon, identifying conclusively the elements which would satisfy the establishment of a genuine link. This is a task which has eluded the international community since the concept was first articulated. Hence there is no objective measure to guage whether the requirement, now expressed in Article 91 of the *LOSC* that there 'exists a genuine link between the state and the ship', has been met.

The genuine link concept is reviewed in this Chapter to illustrate the way in which its inadequacy as a measure of flag State nationality, authority or control has facilitated the growth in IUU fishing. In the absence of any objective reference points for the verification of the genuineness of a grant of nationality, IUU operators have been able to register vessels with flags of convenience thereby avoiding both flag State control and regional obligations that might otherwise be incurred if the flag State was a member of a relevant RFMO. Flags of convenience are examined further below.

The Nationality of Ships

The principle of flag State sovereignty is implicit in Articles 90 and 91 of the *LOSC*. Article 90 provides that, 'every State, whether coastal or land-locked, has the right to sail ships flying its flag on the high seas.' Article 91, *LOSC* addresses the nationality of ships and is extracted in full below.

---

77 H. Meijers, *The Nationality of Ships* (1967) 244-249. Meijers stated that the genuine link has been prescribed for the sole purpose of safeguarding the necessary authority of the flag State in the best possible manner and without this requirement of necessary authority the concept of nationality of ships is largely a fiction and public order at sea is impossible.

78 It is acknowledged that the UNGA requested the Secretary-General to report at the 61st session of the General Assembly on the study undertaken by the IMO into further clarifying the role of the genuine link. UNGA Resolution 59/25, 17 January 2005, paragraph 30.

LOSC, Article 91

1. Every State shall fix the conditions for the grant of its nationality to ships, for the registration of ships in its territory, and for the right to fly its flag. Ships have the nationality of the State whose flag they are entitled to fly. There must exist a genuine link between the State and the ship.

2. Every State shall issue to ships to which it has granted the right to fly its flag documents to that effect.

It follows that a vessel without nationality may be subjected to the jurisdiction of any nation on the high seas.[79] The right of each State to determine the conditions under which it will grant nationality (and to issue documents of registration) is an entrenched principle of international law. The Permanent Court of Arbitration determined in the *Muscat dhows* case[80] that 'generally speaking it belongs to every sovereign to decide to whom he will accord the right to fly his flag and to prescribe the rules governing such grants.'[81]

This historical precedent has been followed by national and international courts. In the 1927 case of the *SS Lotus* the Permanent Court of International Justice (PCIJ) considered an argument put by the French Government asserting that the 'State whose flag is flown has exclusive jurisdiction over everything which occurs on board a merchant ship on the high seas.'[82] The PCIJ affirmed the principle that 'vessels

---

[79] Article 110 of the *LOSC* provides the right of visit by warships, in circumstances where, *inter alia,* the ship is without nationality. This principle has also been upheld in both decisions of domestic courts. See for example the decision of the United Kingdom Privy Council in *Naim-Molvan v Attorney-General for Palestine* 1948 AC 351, in which a stateless vessel was boarded and seized on the high seas by Palestinian authorities. The Privy Council held the vessel's seizure was lawful. See also H.E. Anderson, 'The Nationality of Ships and Flags of Convenience: Economics, Politics and Alternatives' (1996) 21 *Tulane Maritime Law Journal* 139, 141-142.

[80] *Great Britain v France*, Decision of the Permanent Court of Arbitration 8 August 1905. The decision of the Court of Arbitration is reproduced in (1908) 2 *American Journal of International Law* 921-928.

[81] Ibid, 924. The facts of the case involved the granting of a right by France to the Sultan of Muscat to fly the French flag over dhows owned by the Sultan. Britain objected to this on the ground that it was contrary to the Declaration of the 10th March 1862 whereby both France and Great Britain stated they would engage reciprocally to respect the independence of the Sultan of Muscat.

[82] *The Case of the S.S. Lotus (France v Turkey)* [1927] PCIJ (Ser A) No.10, 8. <www.worldcourts.com/pcij/eng/cases/lotuslaw.htm> at 17 January 2004. The facts involved the arrest of a French national (the Master of the *Lotus*) by Turkish authorities following a collision in Turkish waters between the *Lotus* and the Turkish vessel the *Boz-Kourt*. The Turkish vessel was sunk with loss of life. Following representations from the French authorities, the Government of the Turkish Republic agreed to submit to the jurisdiction of the Permanent Court of International Justice.

on the high seas are subject to no authority except that of the State whose flag they fly.'[83] However the PCIJ did link flag State jurisdiction over ships by assimilating the ship to 'the territory of the State the flag of which it flies'.[84] The likening of a ship to a piece of 'floating territory' of the State is not considered today to represent the true basis upon which flag State authority is premised.[85]

In 1953 the US Supreme Court in *Lauritzen v Larsen* confirmed flag State authority and stated:

> Each State under international law may determine for itself the conditions on which it will grant its nationality to a merchant ship, thereby accepting responsibility for it and acquiring authority over it. Nationality is evidenced to the world by the ship's papers and its flag.[86]

The question of the nationality of vessels and the conditions for the grant of nationality was debated by the ILC at length during the early to mid 1950s. As is evident from the examination of the case law above, there is no obligation under international law to place conditions on the grant of nationality to vessels. However the ILC during its work on codifying the law of the sea did examine the basis upon which a grant of nationality might be made.[87] In 1951 special rapporteur Francois submitted to the ILC a list of common principles adopted by a majority of States in determining the nationality of vessels. The 1886 Regulations applying to the use of national flags for merchant ships[88] were of some influence on the rapporteur's Report in 1951. Although rapporteur Francois suggested that the ownership of the vessel be controlled by individuals or companies with a close link with the registering States and that the master of the vessel be a national of the flag State, the ILC rejected the latter requirement.[89] In its 1955 Report to the UN General Assembly, the ILC recommended provisional Articles concerning the Regime of the High Seas. Draft Article 29 as adopted by the ILC in 1955 appears over leaf.

---

83 Ibid.

84 Ibid, 9.

85 Caron, 'Ships, Nationality and Status' above, 400.

86 345 U.S. 571 (1953), 573 and 584. A Danish sailor has instituted an action against a Danish vessel under US law. The US Supreme Court held that Danish law was applicable.

87 Pursuant to UNGA Resolution 899(IX) 14 December 1954, the ILC was requested to devote the necessary time to study, *inter alia*, the regime of the high seas.

88 Adopted by the Institute of International Law Venice, 1896.

89 Yearbook of the ILC (1951) 327-34.

1955 ILC Report to UN General Assembly,
Draft Article 29

Each State may fix the conditions for the registration of ships in its territory and the right to fly its flag. Nevertheless, for the purposes of recognition of its national character by other States, a ship must either:

1. Be the property of the State concerned, or
2. Be more than half owned by
   a. Nationals of or persons legally domiciled in the territory of the State concerned and actually resident there; or
   b. A partnership in which the majority of the partners with personal liability are nationals of or persons legally domiciled in the territory of the State concerned and actually resident there; or
   c. A joint stock company formed under the laws of the State concerned and having its registered office in the territory of that State.

Genuine Link Requirement

The ILC abandoned the work of rapporteur Francois during its eighth session in 1956, and accepted without evidence of detailed debate a proposal submitted by the Netherlands.[90] In its 1956 Report to the UN General Assembly, the ILC stated that 'the criteria it had formulated could not fulfil the aim it has set itself' and further 'that the existing practice in the various States is too divergent to be governed by the few criteria adopted by the Commission.'[91] The draft presented to the UN General Assembly in 1956 was therefore significantly different to that recommended just one year earlier.

[90] However the Netherlands' proposal did note that to establish the presence of a genuine connection account would have to be taken of the ownership of the ship or the nationality of its crew or captain. Yearbook of the ILC (1956) Vol. II, 63. See also, A.G. Oude Elferink, 'The Genuine Link Concept: Time for a Post Mortem?' in I.F. Dekker and H.G. Post (eds) *On the Foundations and Sources of International Law* (2003) 41, 43.

[91] Report of the ILC covering the work of its eight session to the UN General Assembly (1956) included in the Yearbook of the ILC (1956) Vol. II, 253, 279.

ILC 1956 Report to UN General Assembly, Draft Article 29

1. Each State shall fix the conditions for the grant of its nationality to ships, for the registration of ships in its territory, and for the right to fly its flag. Ships have the nationality of the State whose flag they are entitled to fly. Nevertheless, for the purposes of recognition of the national character of the ship by other States, there must exist a genuine link between the State and the ship.

2. A merchant ship's right to fly the flag of a State is evidenced by documents issues by the authorities of the State of the flag.[92]

The concept of a genuine link to test the national character of a vessel can be linked to the ICJ judgement delivered at the time the ILC was deliberating the Draft Articles. *Nottebohm's* case[93] involved a dispute between Guatemala and Liechtenstein over the confiscation of Nottebohm's property by Guatemala. Liechtenstein asserted that Nottebohm was a national of Liechtenstein. The ICJ, whilst agreeing that the grant of nationality was 'within the domestic jurisdiction of the State, which settles, by its own legislation, the rules relating to the acquisition of its nationality; held that the settlement of disputed nationality between two States places the matter on 'the plane of international law.' 'In order to decide this question, preference has been given to the real and effective nationality which is based on stronger factual ties.[94] The ICJ determined that the naturalisation of Nottebohm by Liechtenstein was lacking in the 'genuineness requisite to an act of such importance.'[95]

The ILC commentary to Draft Article 29 indicates that the Commission's views was for the concept of a genuine link between ship and State to be the 'guiding principle' for the recognition of the grant of nationality.[96] The ILC further stated

---

92 Ibid, 259.

93 *Nottebohm's* case , *(Liechtenstein v Guatemala)* [1955] ICJ Rep. 4.

94 Ibid.

95 Ibid.

96 ILC 1956 Report to the UNGA, above, 279. Whilst *Nottebohm's* case was not mentioned in the ILC Sessions on draft Article 5, the proposal by the Netherlands did refer to the case as did the Reference Guide to the Law of the Sea prepared for the ILC by the UN Secretariat. ILC Yearbook, (1956) Vol II, 62. See also, B.A. Boczek, *Flags of Convenience –An International legal study* (1962) 119-120.

that it "wished to make it clear that the grant of its flag to a ship cannot be a mere administrative formality, with no accompanying guarantee that the ship possess a real link with its new State.'[97]

During UNCLOS I, divergent views were expressed on the utility of a genuine link requirement. Whilst the majority of States were in favour of the requirement of a genuine link as an initial premise, there was no unanimity on the criteria against which such a link could be judged.[98] Ultimately, the words 'for the purposes of recognition of the national character of the ship by other States' contained in Draft Article 29 were not included in Article 5 of the *High Seas Convention*. The phrase was reportedly criticised on the ground that it was uncertain, the uncertainty evidently resting upon the words 'genuine link'. Further, the possibility of non-recognition weighed as a source of future international discord. Given that customary international law recognised the right of States to determine the conditions upon which flag status (and hence nationality for the two concepts are linked) will be granted, it seemed querulous to then allow third States the right to withhold recognition of that very nationality. Following an initially unsuccessful proposal by Liberia to remove the right of non-recognition, the plenary session of UNCLOS I voted on the issue and the Liberian proposal won the day. The full text of Article 5 follows.

High Seas Convention 1958,

Article 5

1. Each State shall fix the conditions for the grant of its nationality to ships, for the registration of ships in its territory, and for the right to fly its flag. Ships have the nationality of the State whose flag they are entitled to fly. There must exist a genuine link between the State and the ship; in particular the State must effectively exercise its jurisdiction and control in administrative, technical and social matters over ships flying its flag.

2. Each State shall issue to ships to which it has granted the right to fly its flag documents to that effect.

[97] Ibid.

[98] For a report of the Proceedings at UNCLOS I see, R. Churchill, *The Meaning of the Genuine Link Requirement in relation to the Nationality of Ships* Study prepared for the International Transport Workers Federation (2000) 19.

Attributing some meaning to the Genuine Link Requirement

The cases mentioned above provide an historical context and some guidance on the importance of determining the nationality of ships and the right of States to determine the basis upon which nationality might be granted. However, there is no guidance on what the term 'genuine link' means in practice.

The concept was not extensively discussed during UNCLOS III when the wording in Article 5 of the *High Seas Convention* was adopted. A possible explanation for the limited discussion during UNCLOS III is that the issue had already been tabled for review by the UN Committee on Trade and Development.[99] However, the wording 'in particular the State must effectively exercise its jurisdiction and control in administrative, technical and social matters over ships flying its flag' which was in Article 5 of the *High Seas Convention,* was placed in Article 94 during UNCLOS III. Article 94 lists the duties of flag States whereas the granting of nationality to ships or vessels was dealt with in Article 91.

Some authors have argued that the act of separating the obligation to exercise effective jurisdiction and control, from the right of States to grant nationality, severed the link between the exercise of duties and the existence of a link between vessel and flag State.[100] As it stands under Article 91 of the *LOSC*, a flag State may grant nationality to a vessel subject to there being a genuine link (whatever that may mean). The exercise of flag State duties are a distinct matter with no bearing, it would seem, on the validity of the grant of nationality.

It is not helpful that writers on this point vary greatly in their views on the usefulness of the term genuine link and its meaning.[101] Indeed, an analysis of opinions concluded that 'opinions on these questions are so diverse that it cannot even be said that a predominant view emerges.'[102] However it is argued that under general rules of treaty interpretation, the term genuine link must mean more than just a 'link'. Article 31 of the *Vienna Convention on the Law of Treaties* [103] provides that a 'treaty shall be interpreted in good faith in accordance with the ordinary meaning to be given to the terms of the treaty in their context and in light of its object and purpose.' Hence, in interpreting Article 91 of the *LOSC*, the word 'genuine' must be accorded some meaning.

---

99 See below this section and Elferink, 'The Genuine Link Concept' above, 43.

100 See B. Vukas and D. Vidas, 'Flags of Convenience and High Seas Fishing: The Emergence of a Legal Framework' in O. Schram Stokke *Governing High Seas Fisheries : The Interplay of Global and Regional Regimes* (2001) 53, 61.

101 Churchill, 'The Meaning of the Genuine Link' above, 33-37.

102 Ibid, 37.

103 *Vienna Convention on the Law of Treaties* opened for signature 23 May 1969, 1155 UNTS 331 (entry into force 27 January 1980).

This is the view expressed by Churchill.[104] What that meaning might be remains illusive for there is no agreed criterion against which a genuine link might be judged, and even if there were, there is no consensus on the consequences in the absence of a genuine link.[105] Vukas and Vidas have summarised the position both accurately and eloquently in their conclusion that:

> The search for the elusive 'genuine link' in the law of the sea has eventually led to a 'dead end', not least where fishing vessels are concerned. The increasing number of countries offering so-called 'open registers' and the relaxed instruments and quick procedures to (re-)enter these have made it possible for fishing vessels to 'leave port under one flag, fish under another, and return home under the original.[106]

The term 'genuine link' was given some consideration in the *IMCO* Case,[107] a case which stemmed from a request for an advisory opinion from the ICJ on the meaning of the phrase 'the largest ship-owning nations' as it appeared in Article 28(a) of the *Convention of the Intergovernmental Maritime Consultative Organisation.*[108] A number of States made submissions to the ICJ that the genuine link concept was relevant to determining the meaning of the term 'ship-owning nations.'[109] For example, both Norway and the Netherlands suggested that registration alone did not establish a genuine link.[110] The ICJ reached its opinion without considering the 'genuine link' requirement as it was then stated in Article 5 of the *High Seas Convention*.

---

[104] For example Professor Churchill states that the granting of nationality to a ship by a State by definition creates a link. However, there must be more than just a link or the word 'genuine' would be redundant. Further, the French text of Article 91 of the *LOSC* used the word 'substantiel' and the Spanish text uses 'autentica'. See Churchill, 'The Meaning of the Genuine Link' above, 38-39 and 69.

[105] It is beyond the scope of this study to examine in detail (further than that provided in this Chapter) the meaning of 'genuine link.' There is no normative basis to the term and no enforcement mechanisms in the absence of a genuine link. There are numerous articles in this area of law which discuss the concept. Those mentioned here have been thoroughly reviewed by the author. However for further reading see Churchill, 'The Meaning of the Genuine Link' above; H.E Anderson, 'The Nationality of Ships and Flags of Convenience' above, 139; S.Tache, 'The Nationality of Ships: The Definitional Controversy and Enforcement of Genuine Link' (1982) 16 *International Lawyer* 301; M.S. McDougal, W.T. Burke and I.A Vlasic, 'The Maintenance of Public Order at Sea and The Nationality of Ships' (1960) 25 *American Journal of International Law* 25 and M. McConnell, ' "...Darkening Confusing Mounted upon Darkening Confusion": The Search for the Elusive Genuine Link' (1985) 16 *Journal of Maritime Law and Commerce* 365.

[106] B.Vukas and D.Vidas, 'Flags of Convenience and High Seas Fishing: The Emergence of a Legal Framework', above, 62-62.

[107] *Advisory Opinion on the Constitution of the Maritime Safety Committee of the Inter-Governmental Maritime Consultative Organisation* [1960] ICJ Rep 150. The term 'genuine link' was addressed in several submissions made by intervening States however the ICJ determined that it was not necessary to consider the term in reaching its conclusion. The IMCO was the predecessor of the IMO.

[108] *Convention of the Intergovernmental Maritime Consultative Organisation* opened for signature 6 March 1948 (entry into force 1958).

[109] Elferink, 'The Genuine Link Concept' above, see the text accompanying n 22 of the article.

In 1999 before ITLOS, the concept of a genuine link and the rights of States asserting the absence of a genuine link, were raised by Guinea. The facts of the case arose from the arrest on 28 October 1997 of the oil tanker, the *Saiga*, near the Guinean EEZ.[111] One of the many issues canvassed during both the *M/V Saiga (No 1) Case* and the *M/V Saiga (No 2) Case*[112], was the legitimacy under Article 111 of the *LOSC* of the 'hot pursuit' conducted by the Guinean patrol vessel on the *Saiga*. In this context, ITLOS considered an objection made by Guinea to the admissibility of the application for prompt release by the flag State. Guinea's objection was premised on the submission that no genuine link existed between the *Saiga* and St Vincent and the Grenadines.[113]

Guinea argued that flag state obligations to exercise jurisdiction over registered vessels imposed by Article 94 of the *LOSC* must be fulfilled to establish the 'genuine link' required under Article 91. If this submission were accepted, it was further argued that in the absence of the exercise of any such jurisdiction, no link between the ship and the flag State could exist. Hence, the claims put to ITLOS by St Vincent and the Grenadines as the flag State, could be dismissed on the ground that there was no 'genuine link' between the flag State and the *Saiga*.[114] In essence, Guinea argued that the existence of a 'genuine link' was a prerequisite for the recognition of nationality.

The Tribunal first held that there was no legal basis upon which Guinea could refuse to recognise the right of the *Saiga* to fly the flag of St Vincent and the Grenadines on the ground that there was no genuine link between the ship and the flag State.[115] In doing so the Tribunal referred to the right of any State which 'has clear grounds to believe that proper jurisdiction and control with respect to a ship have not been exercised, to report the facts to the flag State'.[116] Guinea was also unsuccessful in its submission that there was no genuine link between the *Saiga* and St Vincent and the Grenadines. The Tribunal held that Guinea did not produce evidence sufficient to

---

110 See extracts of the submissions by Norway and the Netherlands in L.B.Sohn and J.E Noyes, *Cases and Materials on the Law of the Sea* (2004) 109-113.

111 The *M/V Saiga* was arrested for allegedly smuggling gas oil into Guinean territory. The original case involved an application for prompt release under Article 292 of *UNCLOS* by the flag State. ITLOS determined that the *M/V Saiga* be released upon the posting of a reasonable bond comprising US$400,000 and the value of the gas oil on board when the *Saiga* was arrested: *The M/V Saiga (Saint Vincent and the Grenadines v Guinea)* ('*M/V Saiga (No 1) Case*') [1997] ITLOS judgement dated 4 December 1997, paragraphs 84-85.

112 *Saint Vincent and the Grenadines v Guinea* (*M/V Saiga (No.2) Case)* [1999] ITLOS judgement on the Merits dated 1 July 1999.

113 *M/V Saiga (No 2) Case,* ibid, paragraph 55-75. Guinea also objected to the admissibility of claims on the basis that the *Saiga* was not registered in St Vincent and the Grenadines at the time of arrest and therefore did not have Vincentian nationality at the relevant time.

114 Ibid, paragraph 76.

115 Ibid, paragraph 86.

justify its submission.[117] In a separate opinion, Judge Anderson observed that he did not read paragraph 83 of the Judgement (of the Tribunal) as going so far as to say that the requirement of a 'genuine link', which contains an element of good faith in the word 'genuine' has no relevance at all to the grant of nationality.'[118]

Finally, before examining the consequences stemming from the lack of certainty surrounding the requirements of a genuine link, it is instructive to note the efforts of the UN Committee on Trade and Development (UNCTAD) to develop conditions for the registration of ships.[119] The *United Nations Convention on the Conditions for Registration of Ships*[120] was the result of a UN General Assembly convened conference of plenipotentiaries with the aim of adopting an international agreement outlining the minimum conditions under which ships should be accepted on national shipping registers.[121] In essence the Convention sought to establish a stronger link between flag State and vessel, one that went beyond mere registration. However the Convention has not received widespread support within the community of shipping states (and States with significant merchant fleets) and is unlikely to come into force in the near future.[122]

The first draft of the *Compliance Agreement* also sought to address the problem of re-flagging. It did so by directly requiring flag States to be satisfied first, that there existed a genuine link between it and the proposed flag vessel and second, to determine the existence of that link by reference to 'all relevant factors' including the nationality of the owner of the vessel or where the effective control of the vessel is exercised.[123] However once it became apparent that attempts to:

---

[116] Per Article 94(6) *LOSC*. Although see Churchill, 'The Meaning of the Genuine Link' above, 51 where he states that 'the fact that a State is limited in the action which it can take in respect of a flag State's failure to exercise the various obligations imposed upon it by Article 94, does not mean that no action can be taken where there is no genuine link as required by Article 91.'

[117] *M/V Saiga (No 2) Case*, above, paragraph 87. See also Churchill, 'The Meaning of the Genuine Link' above n 95, 49-50 for discussion of the Tribunal's judgement. Arguably there is support for attributing some meaning to the word 'genuine' however the shipping States have been more successful in their argument that there is no threshold test.

[118] Separate Opinion of Judge Anderson in *M/V Saiga (No.2) case* above, 110.

[119] UNCTAD was acting at the initiative of developing States which wished to clarify conditions of registration, decrease flag of convenience registrations and thereby increase their share in shipping registrations. See, McConnell, 'Darkening Confusion Mounted upon Darkening Confusion' above, 365, 387-9.

[120] *United Nations Convention on the Conditions for Registration of Ships* opened for signature 7 February 1986, 26 ILM 1229. The Convention is not yet in force.

[121] UNGA Resolution 37/209, 20 December 1982. It is even more unlikely to be supported by flag of convenience States involved in fishing.

[122] Article 19 of the Convention requires ratification by 40 States representing at least 25% of world tonnage to enter into force. As of 30 May 2005 there were 12 State parties. See further the comments of Churchill, 'The Meaning of the Genuine Link' above, 61. See also, M.L. McConnell, 'Business as Usual: An Evaluation of the 1986 UN Convention on Conditions for Registration of Ships' (1987) *Journal of Marine and Coastal Law* 435, 449 and Elferink, 'The Genuine Link Concept' above, 48-50 for comments on the Registration Convention.

> define more closely concepts such as the genuine link should exist between a vessel and its flag State, concepts that had defied closer definition in the context of the LOS Convention, would draw the negotiations into a legal quagmire that would bog the whole agreement down indefinitely... the primary focus of the draft agreement was changed from the legal act of flagging and vessel registration to the act of authorising a vessel on the high seas.[124]

In conclusion, the world's courts and leading academics in international law of the sea have been unable to satisfactorily come to terms with the term 'genuine link.' Perhaps the best view of the legacy of that particular ILC Session in 1956 at which the requirement of a genuine link was adopted,[125] is that expressed by McDougal in 1960. He stated that it is most unfortunate that a wholly new provision hitherto unknown either to the customary or treaty law of the sea or to national shipping laws of any particular country, was adopted by the ILC without any attempt to define what is meant by 'genuine link.'[126] For the present purposes it is unnecessary to inquire further as to what connection between the flag State and ship might constitute a genuine link.[127] Indeed it has been argued that pursuing a definition on the international stage is also unnecessary and even unhelpful.[128]

The exercise of reviewing the legal and historical basis for flag State control illustrates two significant legal principles which have had a profound impact on high seas fisheries. First, there are no settled criteria for determining the existence of a genuine link between flag state and ship and second, the grant of nationality to a ship by the flag State is not open to question. In fact the second principle militates against the allocation of yet more international resources to identifying the elements of a genuine link for non-flag States have limited grounds under international law to take any action against the vessel in the absence of a genuine link. In particular, it seems that refusal to recognise a grant of nationality, however tenuous the link may be, would be contrary to international law.

---

123 Elferink, 'The Genuine Link Concept' above, 53 and see G. Moore, 'The Food and Agriculture Organisation of the United Nations *Compliance Agreement*' (1995) 10 *International Journal of Marine and Coastal Law* 412, 413.

124 Moore, ibid, 412, 413.

125 347th meeting of the ILC during the 8th session, 16 May 1956. Yearbook of the ILC (1956) Vol I, 64, 67. The necessity for a genuine link was accepted by the Sub-Committee following the submission of a proposal by the Netherlands.

126 McDougal, 'The Maintenance of Public Order at Sea' above, 104-5.

127 Although the UNGA has invited the IMO to examine and further clarify the role of the genuine link. See UNGA Resolution 59/25 (2005) above.

128 See Stokke and Vidas, 'Regulating IUU Fishing or Combating IUU Operations' above, 8. The authors State that in terms of combating IUU operations at the stage of vessel registration and subsequent licensing, instead of re-opening the eternal discussion about 'genuine link' and 'flags of convenience' one should start by identifying the elements an IUU operator needs to start a voyage. Elferink in 'The Genuine Link Concept' above, refers to the actions of coastal States and port States in securing the compliance of vessels with international rules, particularly in relation to merchant shipping which has parallels for regulating fishing. 53-54 57-58.

The emergence of open State registries, not prohibited at law and driven primarily by economic factors is arguably inevitable. In the next section the increase in flag of convenience registrations is linked to the rise in IUU fishing. In essence, flag of convenience fishing vessels engage in fishing activities free of any flag State control. The practice is reviewed to illustrate the depth of the problem and the need for coastal and ports State action in the absence of effective flag State control.

Flags of Convenience

It has been stated that one of the key issues, if not the key issue in addressing IUU fishing, is the need to achieve more effective flag State control.[129] Whilst flag State rights and duties are not a central focus of this text , an appreciation of the measures to address the reality of demonstrated flag State inaction is required to understand why alternate legal measures such as the ones adopted by *CCAMLR* and Australia are necessary. Therefore the international efforts to improve the level of flag State control exercised over fishing vessels are addressed in the review of international instruments and law in Part 3. Further, whilst the term 'genuine link' is referred to in several instruments, there has been no success in reaching agreement of the constituent elements of the term.

This section focuses on the phenomenon known as 'flags of convenience' as a contributing factor to the prevalence of IUU fishing. With respect to fisheries management, flag of convenience States typically do little to enforce international obligations they have incurred in relation to flagged vessels fishing the high seas and are usually not members of relevant regional fisheries management organisation. For the present purposes it is sufficient to note that the term flag of convenience is well entrenched in maritime law, although it has only comparatively recently been associated with high seas fishing.[130]

Typically, a flag of convenience State is characterised by the failure to exercise responsibility over flagged vessels. Quite apart from implications for fisheries management, lax flag State control impacts vessel and worker safety. The term 'open registry' has also been utilised to refer to States that accept vessel registrations with few conditions. Both terms denote a lack of a genuine link between the vessel and flag State (the only link being the act of registration) indicating the registration is

[129] D. Doulman, 'Illegal, Unreported and Unregulated Fishing: Mandate for an International Plan of Action' (paper prepared for Expert Consultation on IUU Fishing organised by the Government of Australia in cooperation with the FAO, Sydney, 15-19 May 2000) 1 at paragraph 5. <http://www.affa.gov.au/corporate_docs/publications/pdf/fisheries/eciouuf/ausiuu2000.pdf> (12 May 2001).

[130] It is not within the scope of this study to provide a detailed overview of the history of flags of convenience in maritime activities. A brief introduction is provided in Vukas and Vidas, 'Flags of Convenience and High Seas Fishing' above, 54-55. See also, Boczek, *Flags of Convenience,* above, 2.

[131] See the review in Vukas and Vidas 'Flags of Convenience and High Seas Fishing' above, 53. See also R.Carlisle, *Sovereignty for Sale: The Origins and Evolution of the Panamanian and Liberian Flags of Convenience* (1981) Chapter 1.

[132] Boczek, Flags of Convenience, above, 2.

one of mere convenience to avoid incurring international or regional obligations and/or to take advantage of a particular State's inactiveness with respect to the management of fishing vessels.[131] An early definition of flag of convenience formulated in 1962 remains useful today. In his authoritative work entitled *Flags of Convenience*, Boczek defined the terms to mean:

> The flag of any country allowing the registration of foreign – owned and foreign-controlled vessels under conditions which, for whatever reasons, are convenient and opportune for the persons who are registering the vessels.[132]

Other definitions directly acknowledge the economic lure of flags of convenience. In 1983 Bergstrand defined the term as:

> [A] flag of a State whose government sees registration not as a procedure necessary in order to impose sovereignty and hence control over its shipping but as a service which can be sold to foreign shipowners wishing to escape fiscal or other consequences of registration under their own flag.[133]

The attractiveness of flags of convenience to IUU fishers is apparent. States operating such registries generally choose not to comply with international obligations including those arising from the *LOSC*,[134] UNFSA,[135] *Compliance Agreement*[136] and *Code of Conduct* [137] and are not members of relevant RFMOs.

Recent studies of the reliance upon flags of convenience by IUU fishers have concluded that a more accurate description of the service provided by States with open registries is that of providing a 'flag on non-compliance' (FONC).[138] In 2002 *CCAMLR*

---

133 Bergstrand, *Buy the Flag: Developments in the Open Registry Debate* (1983) cited in K.X.Li and J.Wonham, 'Registration of Vessel – New Developments in Ship Registration' (1999) 14(1) *International Journal of Marine and Coastal Law* 137, 141. For further studies and definitions of the term see generally 140-146. In 1970 the UK Committee of Inquiry into Shipping adopted a list of features commonly shared by flag of convenience or open registry States. These included permitting ownership and control of merchant vessels to vest in non-citizens, allowing registration from abroad, law taxes, allowing manning of vessels by non-nationals and a lack of power (be it administrative authority or otherwise) to impose governmental or international obligations upon vessels. (142).

134 *LOSC*, Article 94,

135 *The United Nations Agreement for the Implementation of the Provisions of the United Nations Convention on the Law of the Sea of 10 December 1982 relating to the Conservation and Management of Straddling Fish Stocks and Highly Migratory Fish Stocks*, opened for signature 4 December 1995, 34 ILM 1542, (entry into force 11 December 2001) ('UN Fish Stocks Agreement'), paragraphs 18 -19 and 34-50.

136 *The Compliance Agreement*, Article 1 contains the general duty to ensure flag vessels do not engage in any activity that undermines the effectiveness of international and conservation measures. Arts 2-8 contain additional responsibilities. As of 21 March 2005 there were 29 member states. The Seychelles is one significant flag of convenience state which has ratified the Agreement.

137 *The Code of Conduct*, Articles 6.10, 6.11. Although the duties outlined in the Code are voluntary, it supports obligations in the *Compliance Agreement*, *LOSC* and UN Fish Stocks Agreement.

138 FAO, 'Port State control of Fishing Vessels' FAO Fisheries Circular No. 987. See, also, Schram Stokke and Vidas, 'Regulating IUU Fishing or Combating IUU Operations' above, paragraph 23.

adopted this phrase via Resolution 19/XXI, noting that 'some flag states, particularly certain non-contracting parties, do not comply with their obligations regarding jurisdiction and control according to international law in respect of fishing vessels entitled to fly their flag'.[139] Whilst this new label is undoubtedly more appropriate for RFMOs and international discourse on IUU fishing, it carries no stigma. Flag states disinclined to exercise their general international legal obligations will continue to do so. Stokke and Vidas have proposed adopting the term 'flag of no responsibility' (FONR) to describe those States which fail to exercise any form of control over flagged fishing vessels.[140] The reasoning is that few States would be pleased to be labelled 'irresponsible' in terms of meeting their international obligations.[141]

Whilst this proposal has merit it would seem that some States have already opposed the use of the term flag of convenience (let alone the term FONR) in relation to fishing activities. In 1999, Liberia and Panama, supported by other delegations, opposed the use of the term when the FAO and COFI requested the IMO Sub–committee on Flag State Implementation consider the issue of flags of convenience and their use in fishing activities.[142] It was put to the Sub-Committee that the use of the term was not appropriate and that the term 'open registry States' as used by the IMO was more appropriate.[143] This objection to flags of convenience can be explained in part by the fact that both Liberia and Panama operate significant open registries.

Very recently the role that flag of convenience States play in both facilitating IUU fishing and hindering attempts to deter the practice (by sheltering behind the shield of flag State authority); was highlighted by a comment made by the Australian Federal Minster for Fisheries. In March 2005, six foreign fishing vessels were detected, by Australian authorities, within *CCAMLR* waters just south of the Heard and McDonald Islands' AFZ. The vessels were flagged to Togo and Georgia, both non-member States of *CCAMLR*. Senator Ian MacDonald stated in the context of flag of convenience States, that 'in these flag States the administrative and government arrangements are loose and inadequate and very often corrupt' and that 'international law does not allow any action to be taken meaning there was no recognised right of non-flag State enforcement with respect to high seas fisheries.'[144]

---

139 Resolution 19/XXI (2002) adopted at *CCAMLR*-XXI.

140 Schram Stokke and Vidas, 'Regulating IUU Fishing or Combating IUU Operations' above, paragraph 23.

141 Ibid.

142 M. Hayashi, 'Illegal, Unreported and Unregulated (IUU) Fishing: Global and Regional Responses' in D.D. Caron and H.N. Scheiber (eds) *Bringing New Law to Ocean Waters* (2004) 95, 102-103.

143 Ibid.

144 I. McDonald, Minster for Fisheries, Forestry and Conservation and C. Ellison, Minister for Justice and Customs, 'Flag of Convenience vessels flaunt international rules' (Joint Statement 4 Mar 2005) <www.mffc.gov.au/releases/2005/05028mj.html> (4 March 2005).

Total shipping registrations figures provide an insight to the problem. From 1991–95, 82% of new vessel registrations (fishing or otherwise) were lodged with just 14 nations.[145] More recent figures from the 2002 FAO State of World Fisheries and Aquaculture Report support this trend.[146] Of the vessels reflagged in 2000, 36.3 per cent were flagged to known flag of convenience States.[147] In 2000, the Lloyd's list of shipping indicated over 1000 fishing vessels were registered with Belize, Honduras, Panama, St Vincent and the Grenadines and Equatorial Guinea, all known as flag of convenience States.[148] Other States with open registers include: the Seychelles, Sao Tome and Principe, the Netherlands Antilles, Togo and Russia.[149] Another estimate suggests that up to 1200 flag of convenience fishing vessels arc operating in the high seas without effective flag state control.[150] Estimates by the FAO in 2002 suggest that there are no fewer than 1500 fishing vessels classified as flag of convenience vessels.[151] It is also insightful to consider the comment made in 1999 that 'the open registry as a whole has been growing constantly in terms of number of ships during the last two decades by an average growth rate of 4.19% per annum compared with the world growth rate (in registrations) of only 1.31% per annum.[152]

The variation in the data on flag of convenience registrations sheds light on the nature of the emerging IUU fishing industry and the difficulties created for fisheries

145 The list includes several flag of convenience States. The 14 States are Japan, the European Union, Honduras, Russia, Peru, the former USSR, Chile, Liberia, Morocco, China, Argentina, Iran, South Korea and the US. See, J. Fitzpatrick and C. Newton, Greenpeace International, *Assessment of the World's Fishing Fleet 1991-1999,* 5 May 1998, <http://archive.greenpeace.org/oceans/global overfishing/assessmentfishingfleet.html> (1 October 2004).

146 FAO, *The State of World Fisheries and Aquaculture* (2002) <http://www.fao.org/docrep/005/y7300e/y7300e04.htm#P733_30672> (1 October 2004).

147 Ibid. This figure reflects only those registrations that were able to be tracked.

148 M.Whitfield, 'Plunder of the Seas' (2000) 3 *Transport International* <http://www.itf.org.uk/itfweb/TI/TI3_3.htm> (1 October 2004). The Lloyd's List refers to the shipping List kept by Lloyds Insurance of London.

149 Australian Statement to 23rd Committee on Fisheries, FAO, 15-19 February 2003, 2, <http://www.mffc.gov.au/index.html> (1 October 2004). An open register has been referred to as the situation where foreign vessel owners having no real connection with States maintaining open registers, register their vessels in such states to seek benefits such as avoiding the laws and controls of a state with which they might have a stronger connection. See also, FAO, *Report of the Expert Consultation on Fishing Vessels Operating under Open Registries and Their Impact on Illegal, Unreported and Unregulated Fishing* (2003) FAO Doc FIP/R722. See, K. Rigg, *Halting IUU Fishing: Enforcing International Agreements* April 2004, 17-18 for further detail on flag of convenience State registrations. (prepared for OECD *Workshop on IUU fishing*) above, 369.

150 Antarctic and Southern Ocean Coalition, 'Environmentalists and Trade Union Leaders Have Called for Urgent Action by the Governments to End Pirate Fishing in the World's Oceans' (Press Release, 25 April 1999) <http://www.seastead.org/localres/misc-articles/end_flag_con_04_25_1999.html > at 1 October 2004.

151 J. Swan, Consultant, FAO, *Fishing Vessels Operating under Open Registers and the Exercise of Flag State Responsibilities — Information and Options* (2002) FAO Doc FIPL/C980, 22.

152 K.X. Li and J. Wonham, 'Registration of Vessel – New Developments in Ship Registration' above, 137, 144.

management bodies. The basis for seeking a flag of convenience is that vessel owners wish to avoid both detection and any obligation to comply with conservation and management measures, hence the lack of detail available about such registrations. Adding to the difficulties encountered by coastal states and RFMOs in exercising authority over IUU fishing vessels is the fact that flag of convenience vessels often fly no flag or have the vessel name and number painted over.[153]

It has been observed that in 'a world where strong and effective flag State control was exercised, there would be little scope for IUU fishing.'[154] Whilst this is an accurate observation, there is little economic incentive for States to either decline applications for registration or to take effective enforcement action with respect to non-compliant fishing vessels. Similarly there are significant economic drivers encouraging IUU fishers to seek registration with 'suitable' States and this fuels the market for FONC or FONR States. It should also be noted at this juncture that some developing States lack not the will but the capacity to effectively manage flagged vessels.[155] Under the *LOSC* all States, coastal or land-locked, have the right to sail ships flying their flag on the high seas.[156] The revenue to be gained by developing States, whether through vessel registration or granting port facilities is enticing.

[153] When the *South Tomi* was first detected, she was flying no flag. Interview with John Davis, Australian Fisheries Management Authority (in person 4 June 2001). The *Viarsa* reportedly had a painted out name and vessel number: M. Phillips, 'Fish Poachers Show Colours', *The Herald Sun*, 25 August 2003, 15. The option of boarding the vessel because it is stateless exists under Article 110 of the *LOSC*. This option was considered by Australian authorities during the hot pursuit of the *South Tomi*. Anecdotal evidence is that authorities expressed a strong preference to arresting the vessel for a fisheries offence under Australian law rather than exercising the right of visit under international law or having that right exercised by a third State.

[154] D.J. Doulman, 'A General overview of some aspects of Illegal, Unreported and Unregulated Fishing' (paper prepared for Expert Consultation on IUU Fishing organised by the Government of Australia in cooperation with the FAO, Sydney, 15-19 May 2000) 1 at paragraph 5. <http://www.affa.gov.au/corporate_docs/publications/pdf/fisheries/eciouuf/ausiuu2000.pdf> (12 May 2001). See also FAO, Report of and papers Published at the Expert Consultation of IUU Fishing, FAO Fisheries Report No. 666.

[155] For example States may lack the physical capacity to exert control or the political capacity. In May 2005 two vessels, the *Hammer* and the *Sargo* were detected in the Heard and McDonald Islands AFZ but within *CCAMLR* waters. Both vessels were flagged to *Togo* a FOC State and non-member of *CCAMLR*. Whilst requests where made of the flag State by Australian authorities to board the two vessels, no consent was granted, due no doubt to the existing political and civil unrest in Togo. I. Macdonald, Australian Minister for Fisheries, Forestry and Conservation and C. Ellison, Australian Minister for Justice and Customs, 'Viking Shadows illegal fishermen' (9 May 2005) at <www,mffc.gov.au/releases/2005/05089mj.html> (12 May 2005).

[156] *LOSC*, Article 90.

2.2.6 *Pacta Tertii*

A basic rule of customary international law, *pacta tertiis nec nocent nec prosunt*, dates back to Roman law.[157] This principle is based on the premise of sovereign equality amongst all States. It was codified in the *VCLT* and has been described as a general rule 'so well established that there is no need to cite extensive authority for it'.[158] The principle, as stated in the VCLT, is that '[a] treaty does not create either obligations or rights for a third State without its consent'.[159]

The very nature of international law as a consensual arrangement between States[160] is premised upon this fundamental principle.[161] However, the operation of the *pacta tertiis* rule also exposes a practical weakness in the application of international law. Indeed it has been stated that 'the crux of the dilemma of unregulated fishing lies in the principle of *pacta tertii*.'[162] States may choose not to accept a treaty obligation undertaken by a majority of States, sometimes benefiting from their non-compliance as well as undermining attempts by the international community to regulate State practice. In fisheries management such States are often termed 'free riders.'[163]

In practical terms, this means that regional efforts to manage high seas fisheries can be undermined either by non-compliant third party States or by States who do 'sign up' to the relevant convention but subsequently exercise their right to avoid compliance with selected measures. The application of this rule of law as a facilitator or 'driver' of IUU fishing has also been recognised by Stokke and Vidas who observed in the context of the sovereign equality of States, that '[o]n the high seas… not only practical impediments but also basic legal principles work in favour of IUU operators.'[164] The challenge for fisheries managers is to identify effective measures which respect the rule of law and *pacta tertii*.

---

157 See generally E. Franckx, Pacta Tertiis *and the Agreement for the Implementation of the Provisions of the* United Nations Convention on the Law of the Sea *of 10 December 1982 relating to the Conservation and Management of Straddling Fish Stocks and Highly Migratory Fish Stocks* (2000) 5-7 <http://www.fao.org/legal/prs-ol/po8.pdf> at 1 October 2004.

158 R. Jennings and A. Watts (eds), *Oppenheim's International Law* (9th ed, 1992) 1260-1.

159 *VCLT*, Article 34.

160 International law has been defined as 'that body of law which is composed for its greater part of the principles and rules of conduct which States feel themselves bound to observe, and therefore, do commonly observe in their relations with each other': I. Shearer (ed), *Starke's International Law* (11th ed, 1994) 3.

161 Although not directly relevant to this article, the consensual aspect of international law is also applicable in the development of customary international law. Recognition of the persistent objector doctrine has provided the opportunity for states who object to the development of a rule via state practice to assert they are not bound by that rule once it becomes part of customary international law: *Anglo-Norwegian Fisheries Case (UK v Norway) (Merits)* [1951] ICJ Rep 116, 138-9. See also J. Charney 'The Persistent Objector Rule and the Development of Customary International Law' (1985) 56 *British Yearbook of International Law* 1.

162 Molenaar, 'Regional Fisheries Management Organisations: Issues of Participation" above, 72.

163 Ibid. Molenaar states that the principle can 'tempt' States to ignore commitments made by others and enjoy 'free rider' benefits.'

164 Stokke and Vidas, "Regulating IUU Fishing or Combating IUU Operations' above, paragraph 37.

Summary of Legal Factors contributing to IUU Fishing

The converging application of the legal factors outlined in this Chapter creates a particular environment within which maritime activities are regulated. Rules of international law that governments, fisheries management authorities and scholars apply daily, perhaps without question, in their quest for solutions to the problems created by IUU fishing are in fact hindering their very aim. Fundamental principles of international law which have developed gradually through State practice dictate the activities of all States in maritime areas. With respect to high seas fisheries, the relevant principles are the freedom of the high seas and fishing; the authority of flag States (including the inability to hold flag States accountable to the requirement of a genuine link and the right to refuse to recognise a grant of nationality); and the rule of *pacta tertii.*[165]

Whilst IUU operators do not feel bound to observe relevant international, regional or national obligations and regulations and can quickly alter their practices to adapt to changing economic, policy or environmental factors, RFMOs and coastal State authorities are constrained by rules of international law and do generally feel bound to act in accordance with the rule of law. Further, the swiftness with which IUU operators can change their practices is contrasted with the incremental nature of change in international law. It is not the author's aim to consider the merits of proposing changes to the fundamental principles of international law reviewed in this Part, however it is suggested that there would be little viability in such a proposal.

This work proceeds on the premise that the exigencies of challenges created by IUU fishing dictate that the process of identifying strategies to eliminate and deter IUU fishing be grounded in the current legal environment. Any change to the existing principles reviewed above will be long term and responses to IUU fishing cannot be predicated upon the possibility or probability of such change. The strategies employed by *CCAMLR* and Australia are reviewed in this context. However, in order to identify effective strategies, an appreciation of more than the legal factors facilitating IUU fishing, is required. The social and economic factors which contribute to the situation must also be acknowledged and understood and these are examined in the following Chapter.

[165] See generally Baird, 'IUU Fishing: An Analysis' (2004) above, and see also R. Rayfuse, *Non-flag State Enforcement in High Seas Fisheries* (2004) 17-31.

Chapter 3

Economic and Political Factors influencing the development and practice of High Seas Fisheries and the emergence of IUU Fishing

3.1 Introduction

There are a multitude of non legal factors which contribute jointly and severally to the existence and persistence of IUU fishing. These factors have also been described as 'drivers' of IUU fishing.[1] The factors identified below in this Chapter have been selected following a rigorous review of existing literature on IUU fishing.[2] In particular the OECD *Workshop on IUU Fishing Activities* held in April 2004 generated several papers which directly address the economic and social drivers of IUU fishing.[3]

Political factors are also relevant although more subtle in their impact. However given that the acceptance of international obligations is based on consent, aligning the political agenda of a State with potential treaty obligations cannot be assumed. In the arena of high seas fisheries management, the unwillingness (and to some extent, the incapacity) of States to incur legal obligations does facilitate IUU fishing. Similarly, the unwillingness of States to take action on matters which have a direct economic link to IUU fishing, such as the withdrawal of government subsidies, the implementation of port and market State controls hinders the campaign to deter IUU fishing.[4]

This Chapter proceeds to review the economic and political factors which have arguably contributed to or facilitated IUU Fishing and identifies those factors where action might be taken to lessen their impact.

---

1 For example, see O. Schram Stokke and D. Vidas, 'Regulating IUU Fishing or Combating IUU Operations' paper prepared for the OECD *Workshop on Illegal, Unreported and Unregulated (IUU) Fishing Activities* Paris, 19-20 April 2004. <http://www.oecd.org/document/5/0.2340.en_2649_33901_21007109_1_1_1_1.00.html> Also published in OECD, *Fish Piracy, Combating Illegal, Unreported and Unregulated Fishing* (2004). See also D. Agnew and C. Barnes, *Economic Aspects of IUU Fishing: Building a Framework'* (paper prepared for *OECD Workshop on IUU Fishing*).

2 See n2 to Chapter 2.

3 See generally, OECD, *Fish Piracy* above.

4 The action required of States through the implementation of National Plans of Action under the *IPOA-IUU* is reviewed in Chapter 4.

3.2 Low Risk of Apprehension v. High Economic Returns

It has been noted that the problem of IUU fishing runs much deeper than just the activity on the high seas. There are significant economic incentives which attract transnational investment in what has become a multi-million dollar business. The problem of IUU fishing cannot be completely eradicated until the economic incentives are removed. For example IUU operators in the Southern Ocean work on a simple cost-benefit analysis. The low risk of apprehension, whilst fishing vast tracts of the remote Southern Ocean, is far outweighed by the significant financial gains that may be realised through the sale of a hold of Toothfish. In any event, IUU vessel operators need only avoid coastal States such as France and Australia who operate at a significant disadvantage in conducting patrols far away from their respective mainland territory. Furthermore Member States of *CCAMLR*, the relevant RFMO[5] have no authority to board third states vessels within the *CCAMLR* area of application.[6]

In terms of the profits to be made it has been reported that the illegal trade in Patagonian Toothfish is 'probably more profitable than running drugs or people smuggling.'[7] ISOFISH, an NGO 'established to collect, analyse, verify and disseminate' information on long line fishing in the Southern Ocean, was quoted in 2002 as stating 'probably the easiest way to make a million bucks is to put together a boat and go fishing for a season in the Southern Ocean.'[8] Funds invested in acquiring and equipping a fishing vessel can be recouped in just one fishing expedition.[9]

It is not just the IUU vessel operator who stands to profit from involvement in the trade of IUU sourced fish or fish products. Flag States granting registration to IUU vessels, port and market States involved in landing or trading the catch and the fishing companies, all profit from being involved in IUU fishing. The removal of

---

5 See Chapter 5 for commentary on *CCAMLR*'s role as a RFMO. See also, *CCAMLR*-XVI (1997) Report of the Commission, paragraph 5.5. The Norwegian delegate stated that 'Overfishing, illegal, unregulated and unreported at this moment constitutes a most serious challenge to the reputation and credibility of *CCAMLR* as an intergovernmental organisation for rational management of living marine resources on a sustainable basis'.

6 With the exception of inspection and observation rights amongst member States to *CCAMLR* under Article IX of *CCAMLR*. Further, *CCAMLR* is not a formally declared *UN Fish Stocks Agreement* regional fisheries body and will in all reality probably never be so declared. However, the non flag State rights of boarding and inspection under Article 21 of the Fish Stocks Agreement remain of little practical value. Stokke and Vidas, 'Regulating IUU Fishing or Combating IUU Operations' above.

7 B.Montgomery, 'A Fishy Business', *The Weekend Australian* , 26 October 2002, 28. Interview with D. Carter, Austral Fisheries Chief Executive. Perth, December 2002. Austral Fisheries is licensed to fish within the HIMI EEZ.

8 ISOFISH (International Southern Oceans Longline Fisheries Information Clearing House) was established as a joint venture between conservation organisations and licensed fishing companies in 1997 following the 16th *CCAMLR* meeting. A. Graham was interviewed for the ABC *4 Corners* program *The Toothfish Pirates*. See ABC, 'The Toothfish Pirates' *4 Corners*, 30 September 2002. <www.abc.net.au/4corners/archives/2002b_Monday30September2002.html > (14 October 2003).

9 Ibid.

the economic incentives for flag, port and market States to stay involved in the trade of IUU fish is therefore a crucial step in deterring the occurrence of IUU fishing. Alternatively, intense diplomatic pressure may make the potential economic gains on offer through participation in the IUU fishing industry less attractive.

3.3 Capital Investment and Increases in Technology

Significant capital investment in marine fisheries has resulted in rapid advances in technology and increases in the size of the world fishing fleet. In turn this has meant more fish are being harvested in less time, by a greater number of vessels. The impact of over fishing, excess fleet capacity and over capitalisation in world fisheries was noted in 1995 at the Rome Consensus on World Fisheries.[10] It has been rightly observed that 'the best known example of the impact of human activity on exhaustible marine resources is the pressure that too many ships, too much capital, too efficient harvesting techniques and a lack of care for environmental effects that modern fishing fleets have placed upon fish stocks on all the world's oceans.'[11]

During the late 19th century and 20th century, the fishing industry benefited from great advances in technology. These advances have been described in terms of transforming fishing from a 'trade to a science.'[12] Improvements in technology have application in the marine environment far beyond fisheries. Offshore gas and oil deposits are being accessed far below the water surface and the manganese nodules of the deep sea bed became the focus of international attention once it was shown the technology to harvest them existed.[13]

With respect to fisheries, fishing vessels became mechanised in the late 1800s, enabling trawl nets to be wound in within thirty minutes. Previously it had taken three hours by hand.[14] This development led to nets being recast more frequently, allowing fishers to take advantage of short patches of good weather. Other developments included the introduction of nylon nets, outboard motors, sonar

---

10 Rome Consensus on World Fisheries, 14-15 March 1995.

11 H. Scheiber and D. Caron 'Bringing New Law to Ocean Waters' in Caron and Scheiber (eds) *Bringing New law to Ocean Waters* (2004) 6.

12 R. Baird, Illegal, Unreported and Unregulated Fishing: An Analysis of the Legal, Economic and Historical Factors relevant to its Development and Persistence' (2004) 5(2) *Melbourne Journal of International Law* 299-335. See also, D. Hunter, J. Salzman and D. Zaelke, *International Environmental Law and Policy* (2nd ed, 2002) 674.

13 For example the Report of the UN Secretary General on Oceans and the Law of the Sea (2002) noted that offshore oil exploration had been recorded at a depth of 9,743 feet in the Gulf of Mexico. See generally Part VI, Section B for an overview of developments in the extraction of non-living marine resources and specifically paragraph 231 for details on oil exploration in the Gulf of Mexico. The International Sea Bed Authority was established on 16 November 1994 under the *LOSC* and the *Agreement relating to the Implementation of Part XI of the United Nations Convention on the Law of the Sea* adopted 28 July 1994 (entered into force 28 July 1996).

14 D.H. Cushing, *The Provident Sea* (1988) 258, 296.

and tracking devices, and improvements in communications.[15] Improvements in technology also led to improved boat design and navigational equipment. These improvements meant fishers could travel further afield, in greater safety, to search for new fish populations.[16]

During and immediately after World War II, three significant innovations in technology revolutionised the fishing industry. High powered ships, drag nets and on board freezers transformed fishing vessels into huge factory ships. The ability to process fish on board provided vessel operators with the capacity to freeze the catch at sea thereby enabling voyages to be extended over many months. These improvements in technology significantly improved the cost/benefit ratio of each fishing voyage.[17]

Another significant impact of improved technology is the ability to tranship catches at sea from fishing vessels to larger vessels with onboard processing facilities and freezers. The re-fueling and re-provisioning of fishing vessels at sea is also becoming increasingly common. Advances in navigation techniques such as satellite navigation, enable precise rendezvous points to be pre-determined and regularly used. IUU operators appear to take full advantage of this technology to avoid port State controls. Transshipping IUU catches in remote areas such as the Southern Ocean means that IUU catches can be moved undetected from the vessel which actually harvests the fish. In such circumstances ascertaining the true origin of fish that is placed on transport vessels becomes almost impossible and fish can be readily 'mixed' with legal catches or simply shipped to ports further afield and more willing to receive catches of questionable legality.

Transshipping has been acknowledged as a factor which enables IUU operators to evade and avoid regulation.[18] The *IPOA-IUU* directly acknowledges the significance of transshipping and re-supply which allows not just IUU vessels but all distant water fishing vessels to 'stay at sea for long periods of time, transshipping their catches, refueling, rotating crews, re-supplying bait, food and water.'[19] The

15 Hunter, *et al*, *International Environmental Law and Policy*, above, 674-675. The use of sonar enabled schools of fish to be spotted enabling trawlers to be directed to the target area.

16 J. Jacobson, 'International Fisheries in the Year 2010' (1985) 45 *Louisiana Law Review* 1161, 1164. See also L. Juda, 'Changing Perspectives on the Oceans' in Caron and Scheiber (eds) *Bring New Law to Ocean Waters* (2004) 17, 21.

17 G. Hewison, 'Balancing the freedom of fishing and coastal State Jurisdiction' in E. Hey (ed) *Developments in International Fisheries Law* 1999, 161,166. L. Teclaff, 'Jurisdiction over offshore fisheries- how far into the high seas?' (1996-97) 35 *Fordham Law Review* 409, 412.

18 See the background discussion in K. Rigg, 'Halting IUU Fishing: Enforcing International Agreements' (paper prepared for OECD Workshop on IUU Fishing, above) See also UNGA Resolution 59/25 17 January 2005, paragraph 34.

19 M .Gianni and W. Simpson, 'Flags of Convenience, Transshipment, re-supply and at Sea Infrastructure in relation to IUU Fishing-Management Implications and Recommendations for International action arising from a Case Study prepared for WWF' (paper prepared for OECD *Workshop on IUU Fishing*) above n 2, 11. The relevant paragraphs of the *IPOA-IUU* addressing transhipping are examined in Chapter 4 .

reliance upon transshipping by IUU operators has been directly acknowledged by the UNGA via the statement that, 'common means of conducting illegal, unreported and unregulated fishing involves the unreported or misreported transshipments of fish at sea.'[20]

3.4 The over-exploitation of Living Resources

By 1958 the international community noted that years of unregulated fishing had caused serious depletion in many marine living resource populations. The State parties to the *Convention on Fishing and Conservation of the Living Resources of the High Seas*[21] considered:

> That the development of modern techniques for the exploitation of the living resources of the sea, increasing man's ability to meet the need of the world's expanding population for food, has exposed some of these resources to the danger of being over exploited.[22]

Rather than individually self-regulating to adjust catch levels to account for the rapid advancements in fishing efficiency, high seas fishers have tended to intensify their efforts. This can be explained by reference to the principle of freedom of the seas which has led to the situation where the oceans are owned by no one and managed by no one. The presumption thus arose amongst fishers (and their flag States) that any fish left behind would surely be harvested by other fishers.[23] In essence this is the tragedy of the commons as explained in 1968 by Hardin.[24] Consequently there has been no incentive to manage catch levels to ensure the long term sustainability of the renewable resources, nor to delay or adjust harvests according to fish stock population fluctuations.

With the increases in landed catches[25] facilitated by rapid improvements in technology, far greater quantities of fish were hauled from the oceans than could ever have been imagine by Vattel, a century ago, or even Huxley.[26] Whilst the oceans

---

[20] UNGA Resolution 59/25 17 January 2005, paragraph 34.

[21] Opened for signature 29 April 1958, 559 UNTS 286 (entered into force 20 March 1966).

[22] Ibid, Preamble.

[23] This situation has been described as the 'tragedy of the commons' as discussed above in Chapter 2. See G. Hardin, 'The Tragedy of the Commons' (1968) 162 Science 1234, 1244.

[24] Ibid.

[25] See Graph 1.1 in Chapter 1 for catch statistics from 1948 to 2002.

[26] De Vattel, *The Law of nations (or Principles of the Law of Nations applying to the conduct and affairs of nations and sovereigns)* (1811, 4th ed corrected) 125-126. Vattel observed that fish were common to all and sufficient to supply the needs of all. T.H. Huxley, 'The Herring' (1880-1881) Vol. XXIII *Nature* 607, 612-613. Huxley stated that, notwithstanding the harvesting of 2,500 million herring from the North Sea and Atlantic ocean, there is 'not one particle of evidence that anything man does has an appreciable influence on the stocks of herrings.' The legacy of Huxley's firm views that the oceans were inexhaustible, influenced attitudes towards high seas fisheries management for many decades. See W. Schrank, FAO Fisheries Technical Paper No. 437, *Introducing Fisheries Subsidies* (2003) 27.

might have weathered intensive fishing utilising primitive methods, the systematic harvesting facilitated by new technologies has left fish nowhere to retreat. The impact on non target species including sea birds also increased significantly.[27] The belief that marine living resources were inexhaustible was eventually abandoned when it became apparent that fish stocks could be depleted and could suffer from over-fishing.[28]

3.5 Excess Fleet Capacity

The impact of more efficient harvesting techniques has been compounded by an excess capacity in the world fishing fleet.[29] Dr Diouf, the Director-General of Food and Agriculture organization (FAO), has captured the essence of this threat to marine fisheries in the following statement:

> The Great Oceans are exhaustible. Despite the fact that the majority of all resources are now fully exploited, access to these resources remains open in far too many fisheries around the world. Consequently today there are too many vessels chasing too few fish.[30]

Between 1970 and 1990 the global fishing fleet doubled from 585,000 to 1.2 million commercial fishing vessels.[31] In 1998, a Report compiled by two former FAO Executives on behalf of Greenpeace called for a 50% reduction in the world industrial fishing fleet.[32] This Report refers to large-scale industrial vessels which are defined as being more than 24 meters in length[33] and measuring more than 100

---

27 Longlining in particular, increases the incidence of sea bird mortality. See, eg, *CCAMLR*-III (1984) paragraph 21 where the Commission considered a US report entitled 'Assessment and Avoidance of Incidental Mortality of Antarctic Marine Living Resources' in which is was observed there was 'growing evidence that significant numbers of non-target fish, birds, marine mammals, turtles and other marine organisms are being caught and killed in lost and discarded fishing gear and other debris, as well as caught and killed incidentally during certain fishing operations.'

28 See L. Juda, *International Law and Ocean Use Management. The Evolution of Ocean Governance* (1996) 8-48 for a review of this shift in attitude to the management of fisheries.

29 For an overview of this issue see D. Greboval, 'The International Plan of Action for the Management of Fishing Capacity: Retrospect and Prospect' in M.H. Nordquist and J.M. Moor (eds) *Current Fisheries Issues and the Food and Agriculture* Organisation (2000) 561-582.

30 J. Diouf, FAO Director General, 'Too Many Vessels Chasing Too Few Fish' (Press Release, 1 October 2001) <http://www.fao.org/WAICENT/OIS/PRESS_NE/PRESSENG/2001/pren0158.htm> (1 October 2004). Diouf stated: 'Consequently today there are too many vessels chasing too few fish'.

31 Hunter, *et al*, above, 677.

32 See, J. Fitzpatrick and C. Newton, Greenpeace International, *Assessment of the World's Fishing Fleet 1991–1999,* 5 May 1998, <http://archive.greenpeace.org/oceans/global overfishing/assessmentfishingfleet.html> (1 October 2004).

33 Ibid, The European Union has adopted this measurement. It is also adopted by Article II (2) of the *Compliance Agreement*. Proposed amendments to the *Fisheries Management Act 1990* (Cth) use this measurement as a trigger for increased fines. See *the Fisheries Legislation Amendment (Compliance and Deterrence Measures and other matters) Bill 2003.*

gross tons.[34] Of the estimated 3.5 million fishing vessels worldwide, approximately 38,000 or just 1% met these criteria. Notwithstanding this, they accounted for 50-60% of the total capacity of global fishing vessels.[35] The size of the global fishing fleet has 4.1 million decked and un-decked vessels.[36] Of further concern is the fact that newly constructed fishing vessels are being registered in the first instance with flag of convenience States, raising the presumption that perhaps a proportion of new vessels are built with the specific purpose of being used for IUU fishing activities.[37]

Decommissioning vessels will not fully address the problem of excess fleet capacity, for new technology enables vessels to fish more efficiently, for longer periods and at increasingly remote fishery locations. The international community has addressed the issue at several levels. The FAO *Code of Conduct* recommends that States should prevent over fishing and excess fleet capacity.[38] In 1993 the FAO adopted the *International Plan of Action for the Management of Fishing Capacity*.[39] The OECD High Seas Taskforce has noted with concern that 'excess capacity in fisheries in OECD countries can lead to a spillover of capacity into IUU fishing activities.'[40] It has also been observed that efforts to reduce fleet sizes, for example through the reduction in subsidies, can trigger an increase in IUU fishing activity as the displaced vessels seek alternative fishing areas.[41]

The impact of increasing numbers of industrialised vessels has been catastrophic for many fish species. It is now widely accepted within the scientific community that:

> [E]cological extinction caused by over fishing precedes all other pervasive human disturbance to marine ecosystems, including pollution, the degradation of water quality and climate change.[42]

---

34 Ibid, Lloyds uses gross tonnage as the measure for commercial vessels.

35 Ibid.

36 FAO New Release: 'Countries debate strategies for managing fleet capacity and combating illegal fishing' 1 July 2004.

37 Gianni and Simpson, above, 7. 14% of large scale fishing vessels built in the three years to 2003 were flagged to flag of convenience States. The authors also suggest that the vessels have been purpose built for IUU purposes.

38 *Code of Conduct*, Article 6.3.

39 FAO, *International Plan of Action for the Management of Fishing Capacity* (1999) <http://www.fao.org/docrep/006/x3170e/x3170e04.htm> (1 October 2004) Adopted by COFI and endorsed by the FAO Council in 1999 ('IPOA-Capacity'). See below this section for discussion of the IPOA-Capacity.

40 *OECD Workshop on IUU Fishing,* above.

41 FAO News Release: 'Excess capacity and illegal fishing: challenges to sustainable fisheries' 1 July 2004.

42 J. Jackson, et al, 'Historical overfishing and the recent collapse of coastal ecosystems' (2001) Vol. 293 *Science* 629.

In summary, the advancements in technology and the resulting improved efficiency of vessels, coupled with a greater number of fishing vessels has lead to increases in the recorded catch of global marine fish stocks.[43] In addition excess fleet capacity is linked to over investment in the marine fishing industry via government subsidies.

3.6 Subsidies

Self regulation of the market might ordinarily be expected to correct any excess capacity in the world marine fishing fleet over time such that only those fleet operators that can continue to realise profits would remain in the business of fishing. However, marine fisheries are bedevilled by a practice that States have long adopted to assist their nationals when commercial fisheries become unsustainable. Subsidies can 'artificially elevate profit'[44] such that economic indicators may fail to detect that a particular fishery is no longer economical. As such, 'fishing continues beyond reasonable limits and stocks can be decimated'.[45]

There is general consensus that fishing subsidies do 'stimulate overfishing.'[46] However, it has also been recognised that subsidies can serve a useful purpose in encouraging the development of fisheries and fishing industries in developing regions.[47] The FAO *International Plan of Action for the Management of Fishing Capacity* calls for States to distinguish between beneficial or neutral subsidies and subsidies which have a negative impact on sustainability. Agreement on how the detrimental subsidies may be better managed and potentially eliminated has evaded the international community. First, there is no common ground on what government actions constitute a subsidy and furthermore whether payments or funding to an industry sector associated with the fishing industry are a fishing subsidy.[48]

The negative impact of subsidies on marine fisheries was specifically targeted in the *International Plan of Action for the Management of Fishing Capacity*. Whilst the plan is voluntary, it calls on FAO member states to 'assess the impact of subsidies which contribute to overcapacity'.[49] However as mentioned above efforts to reduce fleet sizes, through the reduction in subsidies, can trigger an increase in IUU fishing activity as displaced vessels seek alternative fishing areas.[50]

---

43 See Chapter 1, Graph 1.1

44 Schrank, *Introducing Fisheries Subsidies* above, 31.

45 Ibid, 32.

46 Ibid, 1.

47 Ibid, 2.

48 Ibid, 2-3.

49 To this end, see FAO Fisheries Department, *Third Ad Hoc Meeting of Intergovernmental Organizations on Work Programmes related to Subsidies in Fisheries* (2003) FIPP/R719, 2.

50 FAO news release: 'Excess capacity and illegal fishing: challenges to sustainable fisheries' 1 July 2004. The FAO Technical Guidelines (No. 9) *Implementation of the IPOA-IUU* (2002), 16, also calls for the termination of subsidies where such support is used or may be used for IUU fishing.

It is not the author's intention to examine in detail the complexities of the economic and international trade considerations to be taken into account in achieving international cooperation and success to minimise the detrimental impacts of subsidies in the fishing industry.[51] Notwithstanding this caveat, it is of importance to acknowledge the significant impact subsidies do have on masking the decline in fish populations and collapse of commercial fishing in particular regions. Whilst excess fishing capacity may be gradually decreased to sustainable levels, decisions on the better management of subsidies may inadvertently contribute to a spill over of legitimate fishers to IUU fishing. In this regard fishing subsidies should be viewed within the broad context of high seas fishing management in the 21st century. Addressing their role in perpetuating over capacity and hence over fishing will require global agreement and action by every State which has institutionalised fishing subsidies.[52] It is not however, simply a matter of withdrawing subsidies from vessel owners, industry and fishers and indeed such action would dramatically increase the incidence of IUU fishing.

With respect to subsidies, all States must cooperate to reduce their fleet through a reduction in subsidies. This may be achieved through both regional and national measures. In addition to unilateral action by States, which will be the most effective action taken, RFMOs can contribute by requiring member States to initiate the gradual reduction of fishing capacity and eliminate negative subsidies.[53]

---

51 For further reading, see Report of the Third ad hoc Meeting of intergovernmental organisations on Work Programmes related to Subsidies in Fisheries, FAO Fisheries Report No. 719 (2003) or R. Grynberg, 'WTO fisheries subsidies negotiations: implications for fisheries access arrangements and sustainable management' (2003) 27 *Marine Policy* 499-511. At 500, Grynberg places the role of subsidies in the wider context of high seas fisheries management and states that subsidies accelerate depletion but that in the absence of property rights in fish, stock depletion will occur in open access fisheries whether there are subsidies or not.

52 See Schrank, 'Fishing Subsidies' above, 14-28 for a review of State practice. Subsidies are utilised in all regions of the world. Norway has a long history of subsidising fishers dating back to 1864 (20-22); despite the fertile adjacent fishing grounds, South American States have at times heavily subsidised domestic fishing industries (24-27) and many European States contribute to their local fishing industry (32).

53 Through the implementation of National Plans of Action for the Management of Fishing Capacity (NPOA-Capacity) under Article 19 of the IPOA-Capacity. Under Article 21 of the IPOA-Capacity, plans were to be adopted by 2002. In 2004 the FAO sponsored the Technical Consultation to review progress and promote the full Implementation of the IPOA to Prevent, Deter and Eliminate IUU Fishing Capacity and the IPOA for the Management of Fishing Capacity. The Report of the Technical Consultation dated May 2004 (TC IUU-CAP/2004/4) found that of the 80 responses received from questionnaires sent to the 188 member States of the FAO only two thirds of responding States had developed or intended to develop a capacity management policy with similar numbers having developed a NPOA-Capacity, 4. (as of December 2003 membership stood at 188 States, see <http://www.fao.org/unfao/govbodies/membernations3_en.asp>)

It is argued that the displacement of the high seas fishing fleet increased fishing efforts on high seas fish populations, resulted in increased competition amongst States and contributed to the emergence of IUU fishing. The need to regulate the new fisheries, such as that of the pollock, became apparent when fish populations collapsed.

3.8 The Entry of Corporate Criminals into IUU Fishing

Allegations that organised criminals have become involved in IUU fishing have now gained credibility and a degree of authority.[62] In May 2005 the Australian Federal Minister for Fisheries, Senator Ian Macdonald stated in opening comments to the Closed Ministerial Roundtable at the *International Conference on Governance of High Seas Fisheries* that, 'IUU fishing is now organised trans-national crime, conducted by national and multinational companies. It is companies who fish, not countries.'[63]

The Australian Minister has been vocal in his condemnation of the involvement of criminals in IUU fishing. In 2003, he stated that, 'it is only through collective action that we can succeed against these organized criminal gangs, and the countries that support them.'[64] This statement echoed similar sentiments expressed three months earlier:

> The IUU fishing operations [are] … not just a fishing boat going down there in the hope of catching a feed. These are large international companies, very, very well organised, and with intelligence almost as good as government intelligence.[65]

Australia is well placed to comment on the entry of corporate criminals into the IUU fishing industry. The Australian EEZ is periodically subject to illegal incursions by foreign fishermen. Australia's northern waters were visited, primarily, by subsistence fishermen, operating traditional small-scale fishing vessels. However reports of sightings and arrests in 2005 indicate that the fishers are becoming

---

62 It is not just NGOs and Governments reporting the criminal element in IUU fishing operations. The FAO acknowledged the high level of sophistication evident in IUU fishing operations in the FAO Technical Guidelines for Responsible Fishing , No.9. (2002), Section 3.2.5. See Baird, 'Corporate Criminals and their Involvement in IUU Fishing: An Australia perspective' (2005) *International Fisheries and Policy Law Review* (in print).

63 I. Macdonald, Minister for Fisheries, Forestry and Conservation, Australia, Australian Opening Statement at (speech delivered at Closed Ministerial Roundtable, Conference on Governance of High Seas Fisheries' St John's Newfoundland and Labrador, Canada, 2 May 2005) <www.mffc.gov.au/speeches/2005/highseasfisheries.html> (13 May 2005).

64 I. McDonald, Minister for Fisheries, Forestry and Conservation, Australia, 'Pirates Beware: Australian front and centre in the fight against illegal fishing' 2 December 2004, <www.affa.gov.gov.au/ministers/macdonald/releases/2003/03166m.html> (22 October 2004).

65 I. McDonald, Minister for Fisheries, Forestry and Conservation, Australia, 'The Howard Government's efforts to deter illegal fishing activities', Address at National Press Club, Canberra, 19 August 2003.

more organised in their forays into Australian waters.[66] The waters adjacent to the Heard and McDonald Islands in the Southern Ocean have been targeted by large-scale, industrial vessels.[67] As the arrest of eight such vessels by the Australian authorities has shown, the IUU vessels in the Southern Ocean are part of complex operations, funded by highly-organised companies, and carry an impressive array of communications and satellite equipment onboard.[68]

Individual and cooperative efforts of States to manage both coastal State and high seas fisheries are being undermined by deliberate strategies adopted by some fishing companies.[69] Deceptive business practices such as the creation of extended and complex corporate arrangement to hamper investigations, repeatedly changing vessel names and call signs and regularly re-flagging vessels are just a few examples of deliberate corporate practice to avoid detection by authorities and to profit from involvement in the IUU trade.[70]

The purpose of including the fact of increasing evidence of organised corporate and criminal involvement in IUU fishing as a factor or 'driver' contributing to IUU fishing activity is to indicate the attractiveness of the legal, economic, social and political environment to IUU operators. The attraction of a low risk, high return activity to organised criminals is self evident. The difficulty created by their involvement is that they will not be deterred by moral imperatives. Indeed, IUU fishing activity has persisted and increased in spite of international, regional and national initiatives adopted since the early 1990s. Making IUU fishing economically unattractive, politically undesirable or legally unacceptable is a challenge for all States.[71]

---

66 See Chapter 8

67 Illegal fishing in northern waters presents different problems, and has resulted in substantial apprehensions. For example, in 2002, 109 vessels apprehended in the AFZ were arrested in the northern portion whilst just two vessels were arrested in the Heard and MacDonald Islands' AFZ. Department of Agriculture, Fisheries and Forestry, Australia, *Overview of Illegal, Unreported and Unregulated Fishing*, <www.affa.gov.au/content/output.cfm?ObjectID=4462332C4-D2A2-4D87-9FE743D39C47AFF9> (31 October 2004). However, the two vessels, the *Lena* and the *Volga*, carried an approximate combined total of 236 tonnes of illegally caught Patagonian Toothfish. The Toothfish seized from the *Volga* was sold by public tender for almost AU$2 million. See ABC News Online, 'Court dismisses poachers' boat appeal', 13 March 2004, <www.abc.net.au/news/newsitems/s1065210.htm> (14 March 2004) and Australian Fisheries Management Authority, 'Operation Sutton – Apprehension of suspected illegal fishing vessel' <www.afma.gov.au/whats%20new/operation%20sutton%20-%20the%20incident.php> (8 April 2004). Notwithstanding the traditional difference in northern and southern AFZ apprehensions, recent reports suggest that IUU operators in northern sections of the AFZ are becoming increasingly organised. An enforcement operation in April 2005 resulted in the apprehension of 29 vessels and 240 illegal foreign fishers by Australian authorities. See I. Macdonald, 'Australian Opening Statement at Closed Ministerial Roundtable, Conference on Governance of High Seas Fisheries' above.

68 This point is substantiated in Part 5 of this thesis.

69 For a comprehensive review of recent practice by IUU operators, see, R. Baird, 'Corporate Criminals and their Involvement in IUU Fishing' above.

70 FAO Guidelines for Responsible Fishing, above, .

71 See, R. Baird, 'Corporate Criminals and their Involvement in IUU Fishing' above.

This challenge is further compounded by the fact that, notwithstanding the limitations of international law, it is aimed at State actors, not individuals or corporations. Thus, whilst the international community considers mechanisms to better improve flag State responsibility,[72] individuals and corporations are exploiting the limitations in international law, taking advantage of flags of convenience, remaining out of the jurisdiction of coastal, port or market States and operating with impunity. Whilst it is not within the scope of this study to consider avenues for assigning individual responsibility to IUU fishers or operators, it is intended that the international framework reviewed in Chapter 4 and the comprehensive review of regional and State practice in the Southern Ocean, will identify mechanisms which remove the economic incentives to IUU fishing. As previously stated, modification to the international rules of law is a lengthy process and in any case the possibility of securing an amendment to the key principles of freedom of fishing and flag State sovereignty on the high seas, is very remote. Thus, even if international law were to be amended to more precisely define the link between flag State, and fishing vessel, it remains applicable to States only. Dealing with individuals operating in a trans-national environment is not a challenge restricted to the law of the sea, it is a challenge for all aspects of international law.[73]

3.9 Summary of Non-legal Factors contributing to IUU Fishing

In summary the economic and political factors driving IUU fishing are equally as important to the legal factors reviewed in Chapter 2. Some factors are immutable. Improvements in technology have dramatically and permanently altered access to fisheries and the environment in which marine fisheries are managed. However, excess fleet capacity and government subsidies of national fishing industries are two contributing factors which can be addressed. The promise of high financial returns with a low exposure to risk in an environment of increased competition amongst legitimate fishers for the dwindling high seas fish stocks is an attractive business proposition for both out of work fishers and corporate entities. The removal of economic incentives is therefore another area which can be addressed. All measures which target the economic factors contributing to IUU fishing need to be applied uniformly and consistently to minimise the risk of IUU operators seeking new fishing grounds in regions where the economic incentives have not been addressed.

---

[72] The international instruments are considered in detail in Chapter 4.

[73] It is noted in this regard that individual responsibility for human rights violations is not a new development. German politicians and military personnel were indicted before the Nuremburg Tribunal after World War II. Most recently, the International Criminal Court was established under the *Rome Statute for the International Criminal Court* opened for signature 17 July 1998 (entered into force 1 July 2002). The Statute creates individual criminal responsibility for acts listed under Article 5 including genocide, crimes against humanity, war crimes and crimes of aggression.

3.10 Conclusion

This Part has examined the complexity of legal and non-legal factors that have contributed to the development of IUU fishing and also impeded the effective management of high seas and to a lesser extent, coastal fisheries. Whilst there has been some work in this area, principally via the OECD *Workshop on IUU fishing*, there has been no comprehensive review of both legal and non-legal factors with the aim of establishing why IUU fishing has been able to develop as an activity and persist. This analysis provides the basis for understanding why the regional and national measures adopted by *CCAMLR* and Australia (discussed in Parts 4 and 5 respectively) have been adopted and how they are taking effect. Prior to exploring these regional and national measures it is necessary to examine the applicable international law which establishes a framework of treaty and customary law within which *CCAMLR* and Australia have formulated their respective policy and legal measures. The next Chapter therefore examines the relevant international instruments, normative standards and developing customary international law applicable to the implementation of legal measures to deter IUU fishing in the Southern Ocean in the absence of effective flag State enforcement.

PART 3

International Measures

FAO *Compliance Agreement* and *IPOA-IUU*,

Requirements for Fishing vessel records

Compliance Agreement:

Article VI.1 calls for the submission to FAO of the following data with respect to vessels authorised to fish on the high seas:

- name of fishing vessel, registration number, previous names (if known), and port of registry;
- previous flag (if any);
- International Radio Call Sign (if any);
- name and address of owner or owners;
- where and when built;
- type of vessel;
- length.

Article VI.2 calls for submission to FAO of the following data with respect to vessels authorised to fish on the high seas, to the extent practicable:

- name and address of operator (manager) or operators (managers) (if any);
- type of fishing method or methods;
- moulded depth;
- beam;
- gross register tonnage;
- power of main engine or engines.

IPOA-IUU:

Paragraph 42 of the *IPOA-IUU* calls on flag States to include all the above data for every vessel in its record of fishing vessels, as well as following additional data:

- the previous names, if any and if known;
- name, address and nationality of the natural or legal person in whose name the vessel is registered;
- name, street address, mailing address and nationality of the natural or legal persons responsible for managing the operations of the vessel;
- name, street address, mailing address and nationality of natural or legal persons with beneficial ownership of the vessel;
- name and ownership history of the vessel, and, where this is known, the history of non-compliance by that vessel, in accordance with national laws, with conservation and management measures or provisions adopted at a national, regional or global level; and
- vessel dimensions, and where appropriate, a photograph, taken at the time of registration or at the conclusion of any more recent structural alterations, showing a side profile view of the vessel.

The Coordinating Working Party on Fisheries Statistics recommends that, where possible, States should indicate whether each vessel listed in the record is actively fishing.

Table 4.1 *Compliance Agreement* and *IPOA-IUU* - Data for Inclusion in Record of Fishing Vessels[32]

[32] This Table is a partial reproduction of Box 2 in the *FAO Technical Guidelines for Responsible Fishing (No. 9) Implementation of the IPOA-IUU* (2002), 27.

In addition, member States are obligated to report promptly to the FAO all relevant information regarding the activities of fishing vessels that have undermined international conservation and management measures. States may also draw to the FAO's attention information on vessels not flying their flag where there are reasonable grounds to believe the vessel has engaged in activity that undermines international conservation and management measures.[33]

The obligation to cooperate in the gathering of information on fishing vessel activity has been identified as a crucial element in deterring IUU fishing[34] and has been adopted by many RFMOs. In particular the *CCAMLR* Commission has established IUU Vessel databases to facilitate the operation of the Catch Documentation Scheme (CDS).[35]

4.2.4 Duty to Cooperate and the role of RFMOs

The duty of State parties to cooperate in the provision to the FAO of information recorded under Article IV has been addressed above. Parties are also charged with the duty to cooperate in the implementation of the *Compliance Agreement* under Article V. A particular and complementary duty to the one to provide information on licensed fishing vessels to the FAO, is the duty to exchange information relating to the activities of fishing vessels sighted and reported to have engaged in activities undermining international conservation and management measures.[36] This information is required to assist the flag State in identifying such fishing vessels. Duties on port States to cooperate with flag States under Article V have been addressed above.

Finally, under Article V of the *Compliance Agreement,* State parties are called upon to enter into, where appropriate, cooperative agreements or arrangements of mutual assistance on a global, regional, subregional or bilateral basis, so as to promote the achievement of the objectives of this Agreement. The reference to cooperating through regional organisations is important because it continues the thematic approach adopted during UNCLOS III and reflected in particular in Articles 118, 63 and 64 of the *LOSC*. Article 118 refers to a duty upon States whose nationals exploit identical living resources or different living resources in the same maritime area, to enter into negotiations for the adoption of measures for the conservation of these living resources and to establish subregional or regional fisheries organisations

---

33 *Compliance Agreement*, Article VI (8).

34 C.C. Schmidt, 'Addressing Illegal, Unreported and Unregulated (IUU) Fishing' (paper prepared for the *OECD Workshop on IUU Fishing Activities*, Paris, 19-20 April 2004, Annex 3, 18. A copy of this paper is available at <http://www.oecd.org/document/28/0,2340,en_2649_33901_23649436_1_1_1_1,00.html > (July 2004). A copy is also in OECD, *Fish Piracy, Combating Illegal Unreported and Unregulated Fishing* (2004).

35 This is examined in detail in Part 4.

36 *Compliance Agreement*, Article V.

to that end. Articles 63 and 64 address the management of straddling and highly migratory fish stocks respectively, and both articles call upon States to cooperate through subregional or regional fisheries organisations. The *Compliance Agreement* reiterates the objective of responsible high seas fishing through cooperative regional arrangements.

4.3 The *UN Fish Stocks Agreement*

The *UN Fish Stocks Agreement* 'applies to the conservation and management of straddling fish stocks and highly migratory fish stocks beyond areas under national jurisdiction.'[37] Whilst the Agreement does not directly address IUU fishing, the approach adopted within for the better management of straddling and highly migratory fish stocks is easily transferable to IUU fishing.[38] Further the State parties to the *UN Fish Stocks Agreement* expressly acknowledge the various problems associated with IUU fishing in the *Agreement*'s Preamble. Such problems include the over utilisation of resources, unregulated fishing, over capitalisation, excessive fleet size, vessels re-flagging to escape controls, insufficiently selective gear, unreliable databases and lack of sufficient cooperation between States.[39]

The management of straddling and highly migratory fish stocks *per se* has not been directly examined in this study for it is the role the *Agreement* plays in the identification of measures to deter IUU fishing that is relevant. Furthermore, whilst the *UN Fish Stocks Agreement* may not be applicable to all incidents of IUU fishing, in the context of the Southern Ocean and specifically the Patagonian Toothfish, the *UN Fish Stocks Agreement* is applicable, for the Toothfish has been identified as a

---

37 *UN Fish Stocks Agreement*, Article 3(1). The exception is that Article 6 and 7 also apply to the conservation and management of such stocks within areas under national jurisdiction, subject to the different legal regimes that apply within areas under national jurisdiction and in areas beyond national jurisdiction as provided for in the Convention.

38 Many authors include the UNFSA within the suite of international instruments that provide guidance in addressing IUU fishing. M.Haward, 'IUU Fishing: Contemporary Practice' in A.G. Oude Elferink and D.R. Rothwell (eds) *Oceans Management in the 21st Century* (2004) 87, 88, identifies the UNFSA as a key instrument to fill lacunae identified in the provisions of the LOS Convention. Further, B.Vukas and D. Vidas, 'Flags of Convenience and High Seas Fishing: The Emergence of a Legal Framework' in Olav S. Stokke (ed) *Governing High Seas Fisheries, The Interplay of Global and Regional Regimes* (2001) 53-90 analyse the provisions of the *UN Fish Stocks Agreement* in the context of the 'recent manifestation of flags of convenience associated with unauthorised fishing in areas of the high seas.'

39 *UN Fish Stocks Agreement*, Preamble. It is noted that selective fishing gear is available and IUU fishers typically do not employ such techniques. *CCAMLR* has adopted conservation measures addressing mesh size, (CM22-01 (1986) and CM22-02 (1984)) and sea bird mortality (CM 24-02 (2004), CM25-01 (1996), CM25-02 (2003) and CM25-03 (2003)). This list of problems reflects some of the non-legal factors identified in Part 2.

straddling stock by *CCAMLR* scientists.[40] Whilst *CCAMLR* is properly described as an organisation concerned with the conservation of Antarctic marine living resources and as such has a much broader scope than typical RFMOs, many factors exist to support its de facto recognition as a RFMO for the purposes of the *UN Fish Stocks Agreement*.[41]

4.3.1 Flag State Controls

Article 18 of the *UN Fish Stocks Agreement* is headed 'flag State duties' and lists a number of measures to be taken by flag States in order to exercise effective control. The approach adopted in the *Compliance Agreement* is followed in Article 18, and States are required to take such measures as are necessary to ensure its flagged vessels fishing on the high seas comply with sub-regional and regional conservation and management measures.

To this end States shall authorise vessels to fly their flag only where they can effectively exercise their flag State responsibilities under the *UN Fish Stocks Agreement* and the *LOSC*.[42] In prescribing measures, such as control via the issuing of fishing licences, the establishment of a national record of authorised fishing vessels; requirements for marking both fishing vessels and gear; regulations for the recording and reporting of vessel position and catch; controls to verifying the catch of target and non-target species; the *UN Fish Stocks Agreement* does significantly expand the nature of flag State duties required under the *LOSC*. [43]

Whilst the *UN Fish Stocks Agreement* applies specifically with respect to the role of flag States in the conservation and management of straddling fish stocks, the approach taken in relation to flag States duties complements that of the *Compliance Agreement* and *Code of Conduct*. The *IPOA-IUU* also calls upon flag States to be more active in managing flagged fishing vessels by requiring all vessel to obtain an authorisation to fish and to maintain vessel records.[44] The cumulative effect of these four international instruments, which represent a combination of hard and

---

40 The *CCAMLR* Scientific Committee raised the possibility of the Toothfsih being a straddling stock in 1992.*CCAMLR*-XIII (1992) *Report of the Thirteenth Meeting of the Commission*, paragraph 143. In 1993 the *CCAMLR* Commission referred to the transboundary distribution pattern of the Toothfish by acknowledging that 'stocks occur both within and outside the Convention Area.' *CCAMLR*-XIV (1993) *Report of the Fourteenth Meeting of the Commission*, paragraphs 57-58. In 1995 the Working Group on Fish Stocks Assessment confirmed that the Toothfish population found in waters adjacent to South Georgia constituted a straddling stock. *CCAMLR*-XVI (1995) *Report of the Sixteenth Meeting of the Commission*, Annex 5- Report of the Scientific Committee, paragraph 10.11.

41 These factors are considered in Part 4.

42 *UN Fish Stocks Agreement*, Article 18(2).

43 For more on flag State duties under the *UN Fish Stocks Agreement*, see, G. Honneland, 'Recent Global Agreements on High Seas Fisheries: Potential Effects on Fisherman Compliance' in O, Schram Stokke, *Governing High Seas Fisheries, The Interplay of Global and Regional Rules* (2001), 121, 134.

44 *IPOA-IUU*, paragraphs 34-50. See below for further detail.

soft international law, is that there are agreed benchmarks for the discharge of flag State duties. Therefore whilst it may be difficult to sustain a case against flag State authority based upon the genuine link requirement, it is foreseeable in the future that action may be instituted against flag of convenience States for failure to discharge the responsibilities prescribed in the *Code of Conduct*, the *Compliance Agreement*, the *UN Fish Stocks Agreement* and *IPOA-IUU*.[45]

4.4 Port State Controls

Article 23(1) of the *UN Fish Stocks Agreement* states that a port State has both the right and the duty, 'to take measures, in accordance with international law, to promote the effectiveness of sub-regional, regional and global conservation'. In particular, the *UN Fish Stocks Agreement* contemplates that port States may 'inspect documents, fishing gear and catch on board fishing vessels, when such vessels are voluntarily in its ports or at its offshore terminals.'[46] Port States may regulate to 'prohibit landings or transhipments where it has been established that the catch has been taken in a manner which undermines the effectiveness of subregional, regional or global conservation and management measures on the high seas.'[47] There is no authority exercisable by port States to detain visiting fishing vessels. Under international law, the sovereignty of port States applies on the basis that visiting vessels are within the territorial jurisdiction of the coastal State.[48] In this case the right of a visiting vessel to depart port would be linked to offences (and corresponding penalties) committed under the port State's national legislation.[49] Potential port States should therefore consider amending national legislation to create appropriate offences linked to involvement in IUU fishing.

4.4.1 Records of Fishing Vessels and Exchange of Information

The requirement of flag States to establish a national record of fishing vessels authorised to fish the high seas is reiterated in the *UN Fish Stocks Agreement* without further elaboration. Members to the *UN Fish Stocks Agreement* are required to provide access to such recorded information on request by directly interested States, subject to any national laws of the flag State regarding the release of such information.[50] In the context of straddling and highly migratory fish stocks, a directly

---

45 As stated previously this would depend upon increased ratifications of the *Compliance Agreement* and *UN Fish Stocks Agreement* and demonstrated State practice in support of these minimum requirements for flag State responsibility.

46 *UN Fish Stocks Agreement*, Article 23(2).

47 *UN Fish Stocks Agreement*, Article 23(3).

48 R.R.Churchill and A.V.Lowe *The Law of the Sea* (1999 3rd ed), 54. This is discussed in detail at further below in this Chapter.

49 Ibid. This legal premise is discussed with approval in T.L. McDorman, 'Regional Port State Control Agreements: Some Issues of International Law' (2000) 5 *Ocean and Coastal law Journal* 207, 222.

50 *UN Fish Stocks Agreement*, Article 18 (3)(c).

interested State might include a coastal State or a State whose nationals are involved in fishing such stocks.

4.4.2 Duty to Cooperate and the role of RFMOs

Article 7(2) of the *UN Fish Stocks Agreement* imposes the general duty to cooperate for the purpose of achieving compatible measures for the management of straddling and highly migratory fish stocks. In giving effect to the duty to cooperate, 'States shall make every effort to agree on compatible conservation and management measures within a reasonable period of time.'[51]

In Part III, the *UN Fish Stocks Agreement* lays down the framework for achieving international cooperation concerning straddling and highly migratory fish stocks. The role of RFMOs as the vehicle through which cooperation may be brought to fruition is emphasised in Articles 8-13. State Parties are obliged under Article 8(1) to pursue cooperation in relation to straddling fish stocks, either directly or through appropriate subregional or regional fisheries management organisations. Further where a subregional or regional fisheries management organisation or arrangement with competence to establish conservation and management measures for particular straddling or highly migratory fish stocks exists, States fishing for such stocks on the high seas and relevant coastal States shall give effect to their duty to cooperate by becoming members of such an organisation or participating in such arrangements or by agreeing to apply the conservation and management measures established by such an organisation or arrangement.[52] Where there is no subregional or regional organisation or arrangement, States shall cooperate to establish such an organisation or enter into an arrangement.[53]

The *UN Fish Stock Agreement* also contemplates cooperation in subregional and regional enforcement in Articles 21 and 22. These articles have been heralded by some authors as an indication that the *UN Fish Stocks Agreement* is breaking new ground in allowing third State boarding and inspection rights of non-flag vessels on high seas areas covered by a subregional or regional fisheries management organisation.[54] In truth the Articles contemplate boarding and inspection on the high seas only of a vessel of another State Party to the *UN Fish Stocks Agreement*. The consent of the flag State to the boarding and inspection regime is gained through ratification of the *Agreement* and thus membership of the relevant subregional or regional RFMO is not required.[55]

---

[51] *UN Fish Stocks Agreement* Article 7(3).

[52] *UN Fish Stocks Agreement*, Article 8(3).

[53] *UN Fish Stocks Agreement*, Article 8(5)

[54] See, eg M.Hayashi, 'The 1995 Agreement on the Conservation and Management of Straddling and Highly Migratory Fish Stocks: Significance for the Law of the Sea' (1995) 29(3) Ocean and Coastal Management 51, 61.

[55] This is also the conclusion of E. Frankx, *Pacta Tertii and the Agreement for the Implementation of the provisions of the United Nations Convention on the law of the sea of 10 December 1982 relating*

4.5 The Application of the *Compliance Agreement* and *UN Fish Stocks Agreement*

Whilst the FAO *Compliance Agreement* is in force, it has been ratified by just 29 States. However, it is the first global agreement to address flag State responsibility[56] and was followed closely by the *UN Fish Stocks Agreement* which reiterates the concept of flag State responsibility. The *UN Fish Stocks Agreement* has been ratified by 52 States which includes 19 individual ratifications by the member States of the European Union and a single ratification lodged by the European Community (EC). The *Compliance Agreement* has a single instrument of ratification lodged by the EC. Allowing for States which have ratified both Agreements (eleven States, including the EC), the total number of individual States subscribing to the concepts of increased flag State responsibility, port State controls, the duty to record information about registered fishing vessels and the duty to cooperate in the exchange of such information and in the functioning of RFMOs, through membership of either the *Compliance Agreement* or *UN Fish Stocks Agreement* rests at 69.[57]

Within this membership, the Seychelles is a notable flag of convenience State. Uruguay, a party to both Agreements, has experienced difficulties recently in exhibiting the requisite degree of flag State control of its flagged fishing vessels.[58] Namibia has experienced difficulties with both port State and flag State control yet has also ratified both Agreements and has become a Commission member to *CCAMLR*. Of the top ten capture fisheries producing States (see Table 4.2 below), only two States, the USA and Norway have ratified both Agreements. High seas fishing States are not demonstrating support for flag State responsibility and until this occurs (if it does) other measures for deterring IUU fishing must be identified.

to the conservation and management of straddling fish stocks and highly migratory fish stocks' FAO Legal Papers Online, No. 8, June 2000, 15-16. Frankx states, 'The consensual nature of international law has been fully respected by Article 21(1). States are only bound by what they freely committed themselves to.

56 Moore, 'The Compliance Agreement', above, 413-414.

57 The individual member States of the European Union have been counted once. See Annex A to this Chapter for a list of the ratifications to 11 May 2005.

58 For example see *CCAMLR*-XXIII (2004) Report of the Twenty-Third Meeting of the Commission, Appendix 3 to Annex V.

State	Compliance Agreement	UN Fish Stocks Agreement
China	not ratified	not ratified
Peru	*23 February 2001*	not ratified
USA	*19 December 1995*	*21 August 1996*
Indonesia	not ratified	not ratified
Japan	*20 June 2000*	not ratified
Chile	*23 January 2004*	not ratified
India	not ratified	not ratified
Russian Federation	*4 August 1997*	not ratified
Thailand	not ratified	not ratified
Norway	*28 December 1994*	*30 December 1996*

Table 4.2 Top Ten producing States (in order of production) for marine and inland capture fisheries in 2002 which have ratified the *Compliance Agreement* and/ or *UN Fish Stocks Agreement*[59]

Increased ratification of both the *Compliance Agreement* and *UN Fish Stocks Agreement* is required. Multilateral treaties are only binding upon and enforceable between State parties.[60] Non-member States are not bound by the increased flag State obligations and although entitled as a matter of customary international law to apply their laws and regulations to foreign vessels voluntarily visiting their ports, are not bound by treaty obligations to enforce port State controls with respect to fishing. Nor are non member states obliged to maintain records of fishing vessels.

The actions taken by the *CCAMLR* Commission are based upon consensus. As evidence of regional State practice, arguably the measures adopted to deter IUU fishing are contributing to the gradual development of international customary law. Thus whilst the limited ratification of both the *Compliance Agreement* and

59 Details of the top ten producing marine and inland capture fishing States can be found in FAO, World Fisheries Report (2004) Article 1, Figure 4, 7. Whilst inland statistics are included the listing is useful as it indicates the lack of support for both the *Compliance Agreement* and the *UN Fish Stocks Agreement* by States heavily involved in capture fisheries.

60 This is an uncontroversial principle of international law. *Vienna Convention on the Law of Treaties* opened for signature 23 May 1969, (1974) UNTS No.2 (entry into force 27 January 1980), Article 34.

the *UN Fish Stocks Agreement* does indicate a lack of broad international support for the application of mechanisms such as port State controls or maintaining vessel registries, this fact does not preclude RFMOs or coastal States (as will be demonstrated in Parts 4 and 5) from implementing the mechanisms. It is through RFMO practice plus increased membership of these two Agreements that customary obligations will emerge. In this regard RFMO member States will need to support the international instruments through ratification. Annex A to this Chapter provides information on the extent to which *CCAMLR* contracting parties which have ratified the *UN Fish Stocks Agreement* and the *Compliance Agreement*.

4.6 *IPOA-IUU*

The *IPOA-IUU* is a voluntary instrument adopted by consensus at COFI on the second of March 2001.[61] It does however address the issues raised above in the *Compliance Agreement*, *UN Fish Stock Agreement* and to a lesser extent in the *Code of Conduct*. Furthermore it does so in the specific context of preventing, deterring and eliminating IUU fishing. These issues are flag State control, port State measures, cooperation, market State measures and the role of RFMOs. The *IPOA-IUU* also envisages State measures such as national legislation and the implementation of market State controls. Whilst the *IPOA-IUU* may be non-binding in nature, it does raise issues of international importance and provides a reference point for RFMOs and national governments seeking guidance on measures to deter IUU fishing.[62]

In 2002 at the World Summit on Sustainable Development in Johannesburg, delegates called for the implementation of the *IPOA-IUU*. An integral part of its successful implementation is the formulation and adoption of National Plans of Action to prevent, deter and eliminate IUU Fishing (NPOA-IUU). To this end the FAO has scheduled a number of workshops in regional areas to assist States in drafting a NPOA-IUU.[63] The Australian NPOA-IUU was presented to COFI in March 2005, the detail of which is examined in Part 5.

---

[61] This position under international law has been affirmed by States lodging declarations or reservations. For example, Mexico is reported as stating at COFI that 'the adoption of the IPAO-IUU does not affect, nor should it be understood as affecting, the rights and obligations of States, in accordance with international law.' W. Edeson, 'The International Plan of Action on Illegal, Unreported and Unregulated Fishing: The Legal Context of a Non-Legally Binding Instrument' (2001) 16(4) *International Journal of Marine and Coastal Law* 603, 608-609.

[62] The *IPOA-IUU* is often referred to as a 'toolbox of measures'.

[63] For example, a Workshop has been scheduled for 1-5 August 2005 in Fiji for Pacific Island States. A further Workshop was scheduled for November 2005 in the African States.

4.6.1 Flag State Controls

The primary obligation on flag States to ensure their flagged vessels do not engage in or support IUU fishing is reiterated in the *IPOA-IUU*.[64] This is to be achieved by ensuring that before fishing vessels are registered, the flag State can exercise its responsibilities in this regard.[65]

The *IPOA-IUU* makes no reference to a genuine link between the flag State and the vessel, rather, the onus is on the flag State to ensure it can exercise its responsibilities irrespective of the degree of connection existing between it and the fishing vessel. In terms of avoiding the difficult debate on the genuine link concept, this is a positive and welcome step. Indeed this was also the position adopted after the first draft of the *Compliance Agreement* when it became apparent that seeking to define more closely the genuine link concept would delay the negotiation process indefinitely.[66] The *UN Fish Stocks Agreement* also adopts this position, focusing on the discharge of flag State responsibilities in preference to prescribing the nature of the link between flag State and vessel.

Further control over fishing vessels is sought through the requirement to obtain fishing authorisations whether the vessel fishes on the high seas or within areas of national jurisdiction.[67] Flag States are to require vessels to carry the authorisations on board.[68] Flag States are also encouraged to ensure that no flagged vessels engage in the re-supply of IUU vessels or tranship IUU fish.[69] The growing problem of 'at sea' re-supply and transhipment can also be addressed despite the lack of flag State control over vessels supporting the IUU industry, through port State and market controls and international cooperation in the exchange of information.

Re-flagging is specifically addressed by urging flag States avoid flagging vessels with a history of non-compliance (subject to some exceptions), deter vessels from re-flagging for the purposes of non-compliance with global, regional or national conservation and management measures, and take all practical steps to prevent 'flag hopping'.[70]

4.6.2 Port State Controls

The *IPOA-IUU* addresses port State controls in more detail than both the *Compliance Agreement* and the *UN Fish Stocks Agreement* although it lacks the binding nature of these two Agreements. The overarching aim of port State

[64] For example, through transhipping or re-fuelling.

[65] *IPOA-IUU*, paragraphs 34 and 35.

[66] Moore, 'The FOA Compliance Agreement' above, 413.

[67] *IPOA-IUU*, paragraph 44.

[68] Ibid, paragraph 46.

[69] Ibid, paragraph 48

[70] Ibid, paragraphs 36, 38, 39.

measures, as outlined in paragraph 52 of the *IPOA-IUU*, is that measures should be implemented 'for port State control of fishing vessels in order to prevent, deter and eliminate IUU fishing.' In implementing such controls, port States should require fishing vessels (and vessels involved in fishing related activities) to provide 'reasonable advance notice of their entry into port, a copy of their authorisation to fish, detail of their fishing trip and quantities of fish on board, with due regard to confidentiality requirements, in order to ascertain whether the vessel has engaged in, or supported IUU fishing.' [71]

In circumstances where there is clear evidence that a vessel granted port access has engaged in IUU fishing activity, the port State should not allow the vessel to land or tranship fish in its port.'[72] The *IPOA-IUU* further calls on port States to develop national, bilateral, multilateral and regional measures for port State control.[73] Paragraph 62 of the *IPOA-IUU* urges States to cooperate to establish 'bilaterally, multilaterally and within relevant regional fisheries management organisations,' compatible measures for port State control of fishing vessels. It was acknowledged during negotiations at the UN Fish Stocks Conference that the refusal of port entry to foreign fishing vessels known to have previously engaged in IUU fishing had already been utilised with a measure of success.[74]

Finally, there is a presumption that fishing vessels flying the flag of non-State parties to a particular RFMO and which have not agreed to cooperate with the particular RFMO, yet are identified as engaging in fishing within the RFMO area of application, may be engaging in IUU fishing.[75] This is similar to the presumption taken in the application of one *CCAMLR* Conservation Measure.[76] The effect of the presumption in terms of port State responsibilities under the *IPOA-IUU* is that the

---

[71] *IPOA-IUU*, paragraph 55.

[72] Ibid. Paragraph 56. The requirement for 'clear evidence' places a burden of proof on port states that could restrict the potential of the *IPOA-IUU* to facilitate far reaching reforms in deterring and eliminating IUU fishing. The *CCAMLR* CDS, discussed below, operates from the premise that all Toothfish imported or exported into the territory of a contracting party or cooperating non-contracting party, must be accompanied by a completed CDS document. Thus there is no requirement for clear evidence of IUU fishing. Incomplete CDS documents trigger a refusal of port access. Furthermore conservation measure 19-07 (2003) 'Scheme to promote compliance by non-contracting party vessels with *CCAMLR* Conservation Measures' raises a presumption of non-compliance in that non-contracting party vessels sighted engaged in fishing activities with in Convention Area are presumed to be undermining Conservation Measures. Conservation Measure 10-02 (2004) obliges contracting parties to licence their flagged vessels meaning unlicensed vessels are presumed to be in contravention of conservation measures.

[73] *IPOA-IUU*, paragraphs 61-62.

[74] Earth Negotiations Bulletin, Summary of UN Fish Stocks Conference, Vol.7 (14) 28 July 1993.

[75] *IPOA-IUU*, paragraph 63.

[76] *CCAMLR*, CM 10-07 (2003), paragraph 3. This measure relates to the presumption that non contracting parties flagged fishing vessels sighted fishing within the *CCAMLR* area of application may be reported to the Standing Committee on Implementation and Compliance (SCIC) for inclusion on the IUU Vessel List.

relevant port State may prohibit landings and the transhipment of catches unless the identified vessel can establish the catch was taken in accordance with the relevant RFMO conservation and management measures.[77]

4.6.3 Records of Vessels and Exchange of Information

Reflecting the legal obligations under the *Compliance Agreement* and the *UN Fish Stocks Agreement*, the *IPOA-IUU* encourages flag States to establish records of flagged fishing vessels.[78] The reality of IUU fishing in which ownership and operational responsibility often rests in two separate entities is acknowledged in the requirement for the recording of details relating to the natural or legal person registering the vessel, operating the vessel and entitled to beneficial ownership. The ownership history of the vessel including any history of non-compliance is also required.[79] Although the recording of this information is dependant on flag State compliance, the expectation that information be recorded and forwarded to the FAO, is an important step in improving international and regional cooperation.[80] Further, it will facilitate the discharge of the requirement to develop and maintain records of vessels fishing within the area of application of RFMOs, including both authorised and IUU vessels.[81]

4.6.4 Duty to Cooperate and the role of RFMOs

The stated objective of the *IPOA-IUU* is:

> To prevent, deter and eliminate IUU fishing by providing all States with comprehensive, effective and transparent measures by which to act, including through appropriate regional fisheries management organisations established in accordance with international law.

Further, to be fully effective, it is envisaged that the *IPOA-IUU* be implemented by all States, flag or otherwise, either directly in cooperation with other States or indirectly through relevant RFMOs.[82] Cooperation, consultation and the sharing of information is identified as an essential element in successfully implementing the *IPOA-IUU* and in deterring IUU fishing.[83]

---

[77] *IPOA-IUU*, paragraph 63.

[78] Ibid, paragraph 42. States are requested to record the information listed in paragraphs 1 and 2 of Article VI of the *Compliance Agreement*.

[79] Ibid.

[80] The *IPOA-IUU* states that flag States should make available to the FAO and as appropriate other States and relevant RFMOs, information about vessels deleted from their records or whose authorisation has been cancelled and to the extent possible, the reasons therefore. *IPOA-IUU*, paragraph 29. Further paragraph 28 requires cooperation amongst States in the exchange of data or information from records of vessels authorised to fish.

[81] Ibid, paragraph 80.5. The role of RFMOs is considered further below.

[82] Ibid, paragraph 9.1.

[83] Ibid.

The focus on State cooperation through RFMOs is again raised in paragraph 28 of the *IPOA-IUU* which requires cooperation in the exchange of information, expertise and technology, in making policies and measures compatible and in monitoring, control and surveillance. RFMOs are then specifically addressed in paragraphs 78-84. States are called upon to give effect to policies and measures adopted by RFMOs whether members or not.[84] In particular States acting through RFMOs are encouraged to consider a number of measures aimed at deterring IUU fishing. Some of these include:

- The development and implementation of comprehensive arrangements for mandatory reporting;
- The establishment and cooperation in the exchange of information on vessels engaged in or supporting IUU fishing;
- The development and maintenance of records of vessels fishing in the area of application of the RFMO including those authorised to fish and those engaged in or supporting IUU fishing;
- The development of methods of compiling and using trade information to monitor IUU fishing;
- The implementation of market-related measures in accordance with this IPOA;
- Agreement on a definition of circumstances in which vessels will be presumed to have engaged in or to have supported IUU fishing; and
- The encouragement of non member States to join the RFMO and fully participate in the RFMO, or at the least participate in and cooperate with the implementation of measures adopted by the RFMO.[85]

4.6.5 Market State or Trade Controls

The *IPOA-IUU* seeks to implement a complementary measure for deterring IUU fishing through market State controls. Whilst recognising the right of States to trade in fish and fish products harvested in a sustainable manner, States are encouraged to take all necessary steps consistent with international law (including international trade law) to prevent fish, caught by vessels identified by the relevant RFMO to have engaged in IUU fishing, being traded or imported into their territories.[86] Furthermore, measures are to be transparent, based on scientific evidence and the use of unilaterally adopted trade-related measures is to be avoided.[87] The use of multi-lateral trade related measures, consistent with WTO requirements is

[84] Ibid, paragraph 79.

[85] Ibid, paragraphs 80 and 83.

[86] *IPOA-IUU*, paragraphs 65 and 66.

[87] Ibid, paragraphs 66 and 67.

recommended.[88] Such multi lateral measures might include catch documentation schemes such as the *CCAMLR* CDS.[89]

4.6.6 Coastal State Obligations

The *IPOA-IUU* identifies and lists a number of ways in which national governments can address IUU fishing activities both unilaterally and via cooperation with other States. Measures specific to coastal States are addressed in paragraph 51 and include:

- effective monitoring, control and surveillance of fishing activities in the EEZ;
- cooperation and the exchange of information with other States and RFMOs;
- requiring every vessel fishing within coastal State waters obtain a valid authorisation from the coastal State;
- entering the authorisation to fish on a record of vessels;
- ensuring at sea transhipment and processing of fish and fish products in coastal State waters is authorised by the coastal State;
- regulating access to its waters; and
- avoiding the grant of licence to a vessel with a history of IUU fishing.

Some of these measures echoe the instruments discussed above. The *Compliance Agreement* places emphasis upon the duty of flag States to grant an authorisation to fish on the high seas and to maintain a record of authorised vessels. . In this instance coastal States are to provide the requisite authorisation to fish within the EEZ and are similarly required to maintain a record of authorisations.

In addition to coastal State measures, the *IPOA-IUU* identifies thirteen separate areas for action by any State whether coastal or not. These are:

- giving full effect to relevant norms of international law;
- ratifying relevant international instruments;
- reviewing national legislation to increase sanctions and remove economic incentives;
- exercising State control over nationals;
- introducing measures to deter non-co-operating States;

88 Ibid, paragraph 69. It is noted in this regard that there have been no challenges under GATT/WTO to multilateral policies adopted to protect the environment and which also impact on free trade. The cases taken to the WTO Panel to date raise issues of unilaterally adopted environmentally based trade controls.

89 The *CCAMLR* CDS is examined in Part 4.

- drafting and adopting National Plans of Action to deter IUU fishing (NPOA-IUU);
- emphasising monitoring, control and surveillance;
- increasing cooperation between States;
- widening publication of IUU activities;
- sharing technical capacity and resources.[90]

It has been stated that no one measure will effectively deter IUU fishing when applied in isolation. Therefore, the identification above of a suite of legal measures which may be implemented by any State, has the potential to contribute to the process of deterring IUU fishing. The extent to which Australia as a coastal State and RFMO member has implemented applicable *IPOA-IUU* measures is examined in Part 5.

4.7 Recent International Meetings addressing IUU Fishing

In the years since the adoption of the *IPOA-IUU* in 2001, there has been several international meetings and conferences aimed at progressing international and regional responses to IUU fishing. As mentioned in Chapter 1, the High Seas Task Force was established in 2003 following the June 2003 meeting of the OECD Round Table on Sustainable Development.[91] The Task Force met for the first time in March 2005. Although it is a high level task force with membership from Canada, the U.K, Australia, Chile, New Zealand and Namibia, significant fishing nations such as China and Spain are not members, nor are any flag of convenience States. The Task Force did however identify a number of action areas from its March 2005 meeting and these areas broadly accord with the subheadings used above when examining the *Code of Conduct*, *Compliance Agreement*, *UN Fish Stocks Agreement* and the *IPOA-IUU*. With the next meeting scheduled for March 2006, the Task Force identified a need to 'elaborate proposals in sufficient detail to enable them to be swiftly implemented'[92] in the following areas:

---

90 *IPOA-IUU*, paragraphs 10-33.

91 The High Seas Taskforce was formally launched on 1 December 2003. <www.high-seas.org> (7 April 2004). See also OECD, 'Ministerial Taskforce to Tackle Illegal, Unregulated and Unreported Fishing' News Release 1 December 2003. <www.oecd.orf/documents/51/0,2340en_1649_201185_20897011_119690_1_1_1 > (7 April 2004)

92 M.Lodge, 'The High Seas Task Force' (paper delivered at the International Conference on the Governance of High Seas Fisheries and the UN Fish Agreement – Moving from Words to Action, St John's Newfoundland and Labrador, 1-5 May 2005, <http://www.dfo-mpo.gc.ca/fgc-cgp/index_archived_e.htm > at 15 May 2005. The author of this paper is in the Secretariat of the High Seas Task Force. See, also, 'First Meeting of the High Seas Taskforce: Summary of Outcomes' 9 March 2005, <www.oecd.org> (May 2005)

- The establishment of an international Monitoring, Control and Surveillance Network through a coalition of like-minded countries;
- The establishment of a global information system on high seas fishing vessels in the form of a publicly available database;
- The development of guidelines for flag State performance such that actual performance of flag States in relation to their high seas vessels could be evaluated;
- The promotion of the responsible port State concept and the strengthening of port State controls;
- Increased control over nationals;
- Improved high seas governance primarily through implementation of international instruments, co-ordination between RFMOs and the harmonisation of measures; and
- Work to enhance public awareness of the social, economic and environmental consequences of IUU fishing on the high seas.

Prior to the High Seas Task Force Meeting, the OECD Committee on Fisheries hosted an international *Workshop on IUU Fishing Activities* with the specific aim of examining IUU fishing from a 'multi-disciplinary approach.'[93] The need for a multi-pronged or layered response to IUU fishing was again acknowledged by delegates at the recent *International Conference on the Governance of High Seas Fisheries and the UN Fish Stocks Agreement – Moving from Words to Action* in May 2005. The next section of this Chapter identifies the significant areas for action, whether sourced in treaty law or under customary international law or undertaken voluntarily, which have been the reference point for action taken by *CCAMLR* and Australia in deterring IUU fishing.

4.8 The State of International Treaty and Customary Law

4.8.1 Acknowledging Flag State Responsibility

The international instruments reviewed above address the pressing need to improve flag State compliance in terms of achieving responsible fishing on the high seas. Continuing to seek methods of encouraging flag States to better manage their

---

[93] OECD, Workshop on IUU Fishing Activities, <http://www.oecd.org/document/28/0,2340,en_2649_33901_23649436_1_1_1_1,00.html >

[94] Indeed, an increasingly frequently expressed comment is that there has been sufficient talk and little action or results in the campaign against IUU fishing. Continuing to devote resources to encouraging States, who have continually demonstrated no inclination to do so, to comply with principles of responsible fishing on the high seas, is not, in the author's view, the best use of international or regional resources.

fishing vessels through a focus on responsibility, rather than the nature of the link between State and vessel is essential. However, many factors militate against further success in the foreseeable future. The authority of flag States, the demonstrated reluctance to fulfil obligations under international law, vagueness surrounding the genuine link concept and the inability of third States to question the genuineness of vessel registration, all operate to give flag of convenience States the upper hand. In addition to these principles of international law, the economic incentives for many developing States to become, and then stay, involved in the business of selling flags of convenience is a factor which complicates the task of identifying a successful solution to the lack of flag State control over fishing vessels.

As stated, the author does not intend to discuss in any detail methods for improving flag State compliance or securing more effective enforcement amongst their flagged fishing vessels.[94] Rather, the common action areas identified and discussed below have been drawn from the international instruments and recent international developments on the basis that their implementation does not rely upon flag State compliance. Success has been achieved by *CCAMLR* and Australia in applying measures which do not rely upon proactive flag State action to deter the incidence of IUU fishing and the legal basis for these measures is now addressed.[95]

4.8.2 Port State controls

The role of port States in deterring IUU fishing is critical. Indeed, if flag States effectively controlled the activities of their fishing vessels, port State control would be largely unnecessary.[96] This has been the experience with the introduction of regional port State controls over the standard of shipping via a series of regional Memoranda of Understanding (MoU).[97]

The concept of control and enforcement by port States in relation to the activities of fishing vessels on the high seas was embraced after States had accepted the principle of port State control with respect to vessel-source marine pollution, and ensuring minimum IMO vessel safety standards.[98] The legal basis for port State control has been expressed in the following terms.

---

95 The degree of success is reviewed in Parts 4 and 5. It is concluded Chapter 9 that these alternative measures must be employed for some time before their impact can be measured in any quantifiable way.

96 This view is shared by Terje Lobach, the Norwegian Minister for Fisheries. See, T.Lobach, 'Advances in Port State Control Measures' (paper prepared for the OECD Workshop on IUU Fishing, above n 93. Also published in OECD, *Fish Piracy* above, 291.

97 Regional port State controls to enforce environmental and safety regulations were a response to the lack of flag State regulation over flagged vessels. The Paris MoU was the first regional MoU signed in 1982. *Memorandum of Understanding on Port State Control in Implementing Agreements on Maritime Safety and Protection of the Environment*, (1982) 21 ILM. 1.

98 O. Schram Stokke and D.Vidas, 'Regulating IUU Fishing or Combating IUU Operations' (paper prepared for the OECD *Workshop on IUU Fishing*, above. Also published in OECD, *Fish Piracy* above, 19.

As a port is part of a State's internal waters over which a State can exercise the same jurisdiction as if the internal waters were part of the land of the State, a foreign vessel in port is subject to the same jurisdiction as an alien on land.[99]

Therefore whilst the flag State does retain authority over a vessel visiting the port of a third State, under international law the sovereignty of the port State prevails whilst the vessel is in port.'[100] The rationale behind port State control for shipping safety, vessel pollution controls and more recently, fishing, is that through the application of port State jurisdiction, obligations in international treaties can be applied against a visiting vessel notwithstanding the fact that the flag State is either not a member of the treaty in question or does not enforce it.[101]

Concerns over the limit of the jurisdictional reach of port States with respect to the acts of fishing vessels can be addressed by stating that the port State applies its national laws and regulations on the basis of acts performed within the jurisdiction of the State. Thus acts such as seeking to unload or transship catches or possibly even the act of arriving in port having engaged in or supported IUU fishing, would trigger the port State's jurisdiction.

In terms of international instruments, the *Code of Conduct* calls on port States to support the achievement of the *Code's* objectives.[102] Whilst there is less emphasis on port State control in the 1993 *Compliance Agreement*, the general duty of port States to monitor visiting fishing vessels for compliance with international conservation and management measures is emphasised.[103] In the event that a vessel is suspected of having engaged in IUU fishing, the port State is required to notify the flag State but is limited in what other action it might take against the fishing vessel.[104] The *UN Fish Stocks Agreement* reiterates and then expands upon, the rights and duties of port States to inspect documents, fishing gear and the catch on board to verify compliance with international, regional and sub-regional measures.[105] If the inspections establish that the catch has origins linked to IUU fishing, the port State may prohibit landings or transshipment.[106] The *UN Fish Stocks Agreement* does not address the detainment of vessel by port States.

---

99 T.L. McDorman, 'Port State Enforcement: A Comment on Article 218 of the Law of the Sea Convention' (1997) 28 *Journal of Maritime Law and Commerce* 305, 308.

100 This position is supported by several authors. See, Churchill and Lowe, *The Law of the Sea* above, 54; McDorman, 'Regional port State Control Agreements' above, 211 and McDorman, ' Port State Enforcement' above, 308.

101 This is in accordance with the rule of international law that treaties do not bind third States. In this instance, the treaty obligations are applied by the member coastal State to a vessel under its jurisdiction by virtue of its presence in port. This view is supported by McDorman, 'Regional Port State Control Agreements', above, 212.

102 *Code of Conduct*, Article 8.3.

103 *Compliance Agreement*, Article V(2).

104 Ibid.

105 *UN Fish Stocks Agreement*, Article 23(1).

106 Ibid, Article 23(2).

The *IPOA-IUU* discusses port State controls and their implementation in some detail. The framework for effective port State controls includes requiring fishing vessels to provide advance notice (and detailed information on their fishing activities) of an intention to enter port.[107] Landing or transshipment is to be denied when there is clear evidence of IUU activity.[108] The achievement of a deterrence effect via port State controls is envisaged through the establishment of compatible measures on a regional or bilateral basis. The *IPOA-IUU* recommends States cooperate to establish 'bilaterally and multi-laterally, and within relevant RFMOs, compatible measures for port State control of fishing vessels.'[109] The recent work by the FAO and IMO on uniform port State measures is consistent with this recommendation.

In 2000 the Joint FAO/IMO Ad Hoc Working Group on IUU Fishing and Related Matters recognised that whilst fishing vessels were not covered by IMO conventions, there was potential to control vessels engaged in the 'transshipment and transport of fish and the resupply of fishing vessels through the application of port state controls similar to those applicable to maritime safety, pollution prevention and living and working conditions.[110] In 2001, the IMO stated that 'cooperation should be extended to the FAO towards developing a port State control regime of its own through a sharing of the IMO's experience and expertise in the matter.'[111]

To further the development of compatible port State measures, the FAO sponsored an Expert Consultation on Port State Measures in 2002 at which a *draft Memorandum of Understanding (MoU) on Port State Measures* (*draft MOU*) was adopted.[112] The *draft MoU* contemplates a number of obligations for port States to achieve a harmonised system of control. It was agreed that any MoU should apply to all vessels engaged in or supporting fishing activities, including not only fishing vessels, but also the transport and logistical support vessels.[113] These obligations are framed in similar terms to those contained in the *IPOA-IUU* and include:

- A requirement that the foreign fishing vessel provide notification in advance of its intention to seek port access. The notification is to be accompanied with details of the vessel's identity, fishing licence, details of the fishing trip and fish on board;

---

107 *IPOA-IUU*, paragraph 55.

108 Ibid, paragraph 56.

109 *IPOA-IUU*, paragraph 62.

110 *Report of the Joint FAO/IMO Working Group on IUU Fishing and related Matters* (2000), Table 3.3.

111 IMO Submission to the UNGA on the implementation of the *UN Fish Stocks Agreement*, 2001.

112 Expert Consultation to review Port State Measures to combat IUU Fishing, Rome 4-6 November 2002. See FAO Fisheries Circular No. 987, *Port State control of foreign fishing vessels'* 2003. The Director General of the FAO convened the Expert Consultation to facilitate the implementation of the *IPOA-IUU*.

113 Lobach, 'Port State Control Measures' above, Annex 1 is a copy of the *Draft MoU* on port State Measures to combat IUU Fishing prepared by the Expert Consultation to Review Port State Measures to combat IUU Fishing, above.

- The obligation to deny landing, transshipping or processing of fish if the vessel which caught the fish is flying the flag of a non-contracting party or non-collaborating party of a RFMO or has been identified as being engaged in, or supporting, IUU fishing activities…unless the vessel can establish the catch was caught in a manner consistent with the relevant RFMO conservation and management measures.[114]
- The obligation on port States to refuse port access to vessels that have engaged in IUU fishing (including known offenders). In this context the members of the Expert Consultation considered that parties to an MoU on port State control may agree to develop IUU vessel lists or FONC lists. Vessels included on such lists should be denied port access.
- Port inspections of vessels and documentation for the purpose of monitoring compliance with relevant conservation and management measures.

The terms of the *draft MoU* were discussed further during the 2004 FAO sponsored Technical Consultation to Review Port State Measures to Combat Illegal Unreported and Unregulated Fishing.[115] Whilst all members attending the Technical Consultation agreed upon the need for port State measures to combat IUU fishing and endorsed the FAO process, reservations were expressed 'about the MoU as a means of implementing the *IPOA-IUU* as it would create a further bureaucratic layer and that its establishment would be a lengthy process.'[116] It was noted that RFMOs have the competence to formulate resolutions and regulations to facilitate port State measures, and that many had already done so to regulate their respective regions.[117]

After discussion and review, the 2004 Technical Consultation agreed to endorse what has been termed a 'Model Scheme', constituting principles and guidelines to be used by States as a reference for the negotiation and adoption of regional MoUs or resolutions. The Report states that:

The provisions contained in the model scheme are for consideration and adoption as appropriate, by States, RFMOs and at the regional level. The model scheme does not affect the competences of RFMOs with respect to port State measures.[118]

---

[114] The *Report of the Expert Consultation* notes that the concept of port State denial has become established within the maritime industry as a mechanism to ensure compliance with the International Safety Management Code (ISM Code) and therefore it may be possible to use a similar means of deterring IUU fishing. Ibid, Section 4.4.

[115] Held in Rome 31 August -2 September 2004.

[116] FAO Fisheries Report, *Report of the Technical Consultation to review Port State Measures to combat Illegal, Unreported and Unregulated Fishing* (2004) No. 759, paragraphs 11-12.

[117] Ibid.

[118] Ibid, Paragraph 17.

The *Model Scheme on Port State Measures to combat IUU Fishing* (*Model Scheme*) adopts minimum requirements for RFMOs or States to use and consult when developing port State measures.[119] The *Model Scheme* provides detailed guidance on the measures to be taken by port States including establishing a checklist of information to be provided by foreign fishing vessels seeking port access and instructions for both the training of port State inspectors and the conduct of port inspections.[120]

Notwithstanding the FAO sponsored *draft MoU* adopted in 2002 and *model Scheme* agreed upon in 2004, the only source of obligation to apply port State controls stems from regional fisheries conventions such as *CCAMLR*. Whilst the FAO initiatives on port State control as means to curb and deter IUU fishing are encouraging, they have no status in international law except to provide evidence of emerging State practice in the sphere of fisheries related port State controls. It is the author's view that there is currently no obligation under customary international law to apply port State controls with respect to visiting fishing vessels.

The contrary view, based upon the presence of the requirement to apply port State controls in the *Compliance Agreement* and *UN Fish Stocks Agreement,* has been expressed.[121] In response to this contrary view it is submitted that the *Compliance Agreement* and *UN Fish Stocks Agreement* do not yet enjoy the broad international support required to transform treaty obligations into broader customary obligations. Ratifications remain low[122] and important distant water fishing nations such as Japan, China and South Korea have not ratified.[123] Further support via increased ratifications and demonstrated State practice through RFMOs and individual State action is required before a case can be put that the duty to impose port State controls on fishing vessels has emerged as a customary international law duty.

4.8.3 Records of Vessels and Exchange of Information

The establishment of registries of authorised fishing vessels, whether through the FAO, via RFMOs or nationally, is a common theme in the international instruments reviewed above. It is a central pillar in building a multi-disciplined response to

119 Ibid, Paragraph 18. Appendix E to the *Report of the Technical Consultation to review Port State Measures to combat Illegal, Unreported and Unregulated Fishing* provides a copy of the *Model Scheme.*

120 Ibid, Annexes A-E to Appendix E.

121 See the FAO *Technical Guidelines for Responsible Fisheries No. 9 (Implementation of the IPOA-IUU)* 2002, 39.

122 See Annex A to this Chapter.

123 Japan and South Korea have ratified the *Compliance Agreement* but not the *UN Fish Stocks Agreement*. China has ratified neither. This view that the Agreements lack broad international support is endorsed by Molenaar, 'New Areas and Gaps' above, 9 and 18. Molenaar states that the port State scheme remains voluntary and that implementation through a third Annex to the *UN Fish Stocks Agreement* may be worth consideration.

IUU fishing. Concerns in relation to the safeguarding of the information and also controlling access to stored information, need to be addressed, in addition to settling clear criteria for listing vessels identified as IUU vessels.

There is no one comprehensive global database for recording fishing vessels details. The FAO has observed that this fact 'undoubtedly creates opportunities for IUU vessels to escape detection.'[124] The HSVAR established by the FAO is intended to be a repository for information on all vessels registered and authorised to fish the high seas. As such, its function is to provide access to information about, but not limited to, vessel specifications, the name of the vessel, the flag State and previous flag, details of the vessel's owner and operator and the fishing method authorised to be used.

There are currently over 5,500 vessels recorded on the FAO HSVAR representing information from only 17 States. The potential of the HSVAR is therefore yet to be realised. Moreover the HSVAR includes *all* vessels authorised to fish on the high seas thus the distinction between legal vessels and IUU vessels may be difficult to deduce. Rather than duplicating the work already undertaken by RFMOs in compiling IUU vessels lists covering vessels active within their particular area of geographical coverage, the FAO could act as a central repository for IUU vessel data to establish a centrally accessed database. Provided care is taken to safeguard any commercial confidentiality and access is limited to authorised personnel, a central database has the potential to enhance the ability of national and regional authorities to take effective action to deter IUU fishing.

To be effective access should be given to all available data in preference to corralling information into RFMO sectors. IUU operators and fishers are mobile and move freely between RFMO areas of application and hence RFMO member states need to know what IUU vessels are active throughout the world's oceans. Restricting the focus only to IUU vessels in a particular RFMO area is counterproductive particularly in light of the duty to cooperate in the management of high seas fisheries.

4.8.4 Duty to Cooperate and the role of RFMOs

The duty to cooperate in the management of high seas fisheries resources is emphasised in the *LOSC*,[125] the *Compliance Agreement*,[126] the *UN Fish Stocks Agreement*,[127] and the *IPOA-IUU*.[128] The argument that this duty to cooperate,

[124] FAO *Technical Guidelines for Responsible Fisheries No. 9 (Implementation of the IPOA-IUU)* 2002, above, 26.

[125] *LOSC*, Article 63-64 and 118.

[126] *Compliance Agreement*, Article V.

[127] *UN Fish Stocks Agreement, Article* 7(2).

[128] *IPOA-IUU*, paragraph 9.1.

through the medium of RFMOs, has evolved from a treaty obligation to become a matter of customary international law is well grounded. It has recently been stated 'that it is arguable that States may now accept, as a matter of customary international law, the obligation to cooperate in respect of high seas fisheries through the medium of RFOs or other cooperative arrangements.'[129] Prior to this in 1994, Burke observed that the obligation 'is probably considered also to be a customary law of the sea principle.'[130] In support of this statement, he referred to the ICJ decision in the *Fisheries Jurisdiction Case*,[131] in which the ICJ found that the parties 'were required to examine together in light of the scientific and other available information, the measures required for the conservation and development and equitable exploitation' of resources in the high seas beyond 12 miles.'

The duty to cooperate under the *LOCS* was judicially considered and affirmed by ITLOS in the *Southern Bluefin Tuna cases*.[132] This case involved an ongoing dispute between Australia and New Zealand with Japan[133] over the limit set for the total allocated catch and the legality of Japan's experimental fishing program to test the recovery of the tuna stocks at various places.[134] Of relevance to this study, Australia and New Zealand argued that Articles 64, and 117-119 of the *LOSC* imposed a duty on Japan to cooperate in the conservation and management of high seas resources.[135] In granting Provisional Measures, ITLOS considered that there was a duty to cooperate directly or through appropriate international organisations with a view to ensuring conservation and promoting the objective of optimum utilisation of highly migratory species.[136] Further, ITLOS considered that the conduct of the member parties to the *Commission for Conservation of Southern Bluefin Tuna* as

---

129 R.Rayfuse, ' To our Children's Children's Children: From Promoting to Achieving Compliance in High Seas Fisheries' (paper presented to International Conference on the Governance of High Seas Fisheries and the United Nations Fish Agreement, St John's, Newfoundland and Labrador, 1-5 May 2005, 10). <www.dfo-mpo.gc.ca/fgc-cgp/documents/rayfuse_e.htm> (14 May 2005).

130 W.T.Burke, *The New International Law of Fisheries: UNLOS 1982 and Beyond* (1994) 121-122.

131 *UK v Iceland* (1974) ICJ Rep. 31.

132 *New Zealand v. Japan* and *Australian v. Japan* (*Southern Bluefin Tuna Cases*) (Order for Provisional Measures) ITLOS Case Nos. 3 and 4, 27 August 1999.

133 This dispute between Australia and New Zealand, and Japan arose out the unilateral declaration by Japan of a three year experimental fishing program for Southern Bluefin tuna setting a total allowable catch of 1464 tonnes per annum. For the background to the dispute including the formation of the 1993 *Convention for the Conservation of Southern Bluefin Tuna* see C. Romano, 'The Southern Bluefin Tuna Dispute: Hints of a World to Come…Like it or Not' (2001) 32 *Ocean Development and International Law* 313, 315-317 and B, Kwaitkowska 'The Australian and New Zealand v. Japan Southern Bluefin Tuna (Jurisdiction and Admissibility) Award of the First Law of the Sea Convention Annex VII Arbitral Tribunal' 16(2) (2001) *International Journal of Marine and Coastal Law* 239. The Southern Bluefin Tuna was managed under the *Convention for the Conservation of Southern Bluefin Tuna* 1993, opened for signature 1993, 26 Law of the Sea Bulletin 57 (entered into force on 20 May 1994).

134 Romano, 'The southern Bluefin Tuna Dispute' Ibid, 316.

135 Australian and New Zealand Statement of Claim, 15 July 1999.

136 *Southern Bluefin Tuna Cases*, above, paragraph 48.

Introduction

This Part examines and evaluates measures adopted by the *CCAMLR* Commission, and implemented by *CCAMLR* contracting parties, with the aim of deterring IUU fishing in the Southern Ocean. As discussed in Part 3, the duty to cooperate in the management of high seas fisheries has been established as a matter of customary international law. RFMOs are identified as the primary mechanism through which such cooperation should be pursued. It is through regional (and sub-regional) cooperation that measures to deter IUU fishing such as port and market State controls, vessel rccords (IUU and licensed) and the exchange of information, can be employed most effectively.

Part 2, Chapter 2, identified a number of international legal principles which operate to restrict the effective regulation of high seas fisheries on a regional basis when States chose not to comply with relevant conservation measures. These include: flag State authority, freedom of fishing on the high seas and the rule of *pacta tertii*. The successful implementation of *CCAMLR* conservation measures depends not only upon contracting party compliance (which has not always been forthcoming) but also non-contracting party (third State) compliance. Under existing international law, third States cannot be compelled to accept treaty obligations[1] and under the existing law of the sea, all States may freely exercise the freedom to fish the high seas and exert flag State authority over their flagged fishing vessels.

Whilst the measures reviewed in this Part do require contracting party implementation and the cooperation of non-contracting parties, they work independently of the principles of freedom of fishing on the high seas and flag State authority. Therefore the measures are not weakened or undermined because flag States do not comply, even if a particular flag State is a contracting party. This is because the measures are grounded in alternative sources of international law.[2] For example port State controls rely upon the principle of port State jurisdiction as reviewed in the previous Chapter. Should the flag State fail to exercise the required control over its vessel, port States can exert control at the point of landing. Thus the measures seek to influence the behaviour of the IUU operator or fisher by increasing the difficulties encountered when the IUU fish is unloaded or sold.

The rule of *pacta tertii* is also observed in relation to the consent of the port State in choosing to apply port State controls. However, the Achilles heel of high seas fisheries management, that of the freedom of fishing and to a lesser extent flag State authority does not bear the full weight of the expectation of the international community. That is, the successful management of our high seas fish stocks for

1 *Vienna Convention on the Law of Treaties (VCLT)*, opened for signature 23 May 1969, 1155 UNTS 331 (entered into force 27 January 1980), Article 34.

2 On this basis the Vessel Monitoring System (VMS) implemented by the *CCAMLR* Commission is not examined as it relies upon flag State consent to implement, except to the extent that the VMS data can be required by port states to support the verification of documents submitted under the *CCAMLR* CDS.

current and future generations, does not balance on the uncertainties of responsible fishing and the discharge of flag State duties. In this respect the initiatives of the *CCAMLR* Commission (which represents but one RFMO) are valuable topics of study.

Before *CCAMLR* action is reviewed, it is necessary to examine the structure of *CCAMLR*. Chapter 5 reviews the formulation and adoption of conservation measures and the application of international, regional and national legal regimes in the Southern Ocean. The status of *CCAMLR* as a RFMO notwithstanding the Convention's stated objective as a conservation organisation, is also considered. Chapter 6 then reviews in detail the CCALMR response through conservation measures and resolutions.

Chapter 5

CCAMLR Management in the Southern Ocean

5.1 An Introduction to the *CCAMLR* structure

Against a background of over fishing of Antarctic marine living resources, negotiations began in the 1970s, between *Antarctic Treaty* Parties, for a Southern Ocean management regime. The concept was discussed at *Antarctic Treaty* Consultative Meetings in 1975 and 1977,[3] followed by a special conference of parties in May 1980. The text of *CCAMLR* was adopted in Canberra, Australia, on 20 May 1980. The Convention entered into force on 7 April 1982 following receipt of the eighth instrument of ratification.[4]

A total of 32 State parties are united under *CCAMLR*. At the close of 2004, Commission membership stood at 24 States,[5] with another eight (8) States having ratified the Convention without becoming members of the Commission.[6] The newest Commission member, Namibia, joined as a result of diplomatic approaches by both the *CCAMLR* Commission and individual member States.[7]

The *CCAMLR* Commission was established under Article VII and membership is open, *inter alia,* to each contracting party which participated in the meeting at which the Convention was adopted. Acceding States may also qualify for Commission membership. States may accede to the Convention under Article XXIX if they are 'interested in research or harvesting activities in relation to the marine living resources to which this Convention applies.' Entitlement to Commission membership requires a further degree of participation in the Antarctic ecosystem. Article VII(2)(b) provides that:

> [E]ach State Party which has acceded to this Convention pursuant to Article XXIX shall be entitled to be Member of the Commission during such time as that acceding Party is engaged in research or

---

3 ATCM VIII 1975, Recommendation VIII-10; ATCM IX 1977, Recommendation IX-12

4 *CCAMLR,* Article XXVIII.

5 Members: Argentina*, Australia*, Belgium* Brazil, Chile*, EU, France*, Germany, India, Italy, Japan*, Republic of Korea, Namibia (joined 2001) New Zealand*, Norway*, Poland, Russian Federation,* South Africa*, Spain, Sweden, Ukraine, UK*, USA*, Uruguay. The * denotes one of the 12 original Antarctic Treaty members, including the then USSR.

6 The acceding States are Bulgaria, Canada, Finland, Greece, Mauritius, the Netherlands, Peru and Vanuatu.

7 *CCAMLR*-XVIII (1998) Report of the Seventeenth Meeting of the Commission, paragraph 1.7.

> harvesting activities in relation to marine living resources to which this Convention applies.

Thus the hurdle for accession is simply an *interest* in research or harvesting, whereas the benchmark for Commission membership is *actual* engagement in research or harvesting. In either case, a State with conservation interests, as distinct from commercial interests, may qualify for both accession to *CCAMLR* and Commission membership, on the basis of its research interests and activities.

Shortly after entry into force, the Convention faced its first potential test with the development of the krill fishery. Krill is central to the Antarctic ecosystem and there was concern that overexploitation of krill could potentially lead to the collapse of many dependent species.[8] However, difficulties with processing the krill catches and the expanding fin fish fishery averted attention from the krill fishery,[9] and in doing so, avoided for the time being, a test of the *CCAMLR* Commission's authority.

In April 2002, the 20th anniversary of the coming into force of *CCAMLR* was heralded as a:

> [L]andmark in the process initiated by the Antarctic Treaty Consultative Parties, of comprehensive and systematic protection of the Antarctic environment and dependent ecosystems from harmful human interference.[10]

However it was noted at the same time that:

> The illegal, unregulated and unreported fishing…of Toothfish is one of the greatest challenges the Commission has had to face. In recent years the catch rate of illegal, unregulated and unreported fishing has been more than double that of the regulated fisheries.'[11]

IUU fishing for the Patagonian Toothfish has presented a real challenge for the Commission. Responses since the problem was officially acknowledged as an agenda item in 1997 by the Commission[12] have been criticised as being reactive by several NGOs, the media and the legal fishing industry.[13] Much of this criticism has been levelled without knowledge of the complex factors involved in *CCAMLR*

[8] K.H. Kock, 'Fishing and conservation in the Southern Ocean' (1994) 30(72) Polar *Record* 3, 5.

[9] Ibid.

[10] Statement for the Commemoration of 20 years of *CCAMLR*. Annexed to *CCAMLR*-XX (2001) Report of the Twentieth Meeting of the Commission.

[11] Ibid.

[12] *CCAMLR* –XVI (1997) Report of Standing Committee on Observation and Inspection, paragraph 1.1.

[13] For example, Greenpeace has conducted press conferences at the *CCAMLR* annual meetings where it rebukes the Commission for failing to stop IUU fishing. The Media has also joined the debate with several articles reporting less than favourably on *CCAMLR*'s achievements. See, B. Montgomery, 'A Fishy Business', *The Weekend Australian,* 26-27 October 2002, 28; The *Mercury,* 'Snag in Toothfish war' 29 October 2003, 16 and 'Editorial' 8 November 2003, 12. Finally, the Coalition of Legal Toothfish Operators (COLTO) was established in March 2003 reportedly because of 'frustration

decision making[14] or an appreciation of the conflicting political, legal and economic issues operating within international fisheries management.

5.2 Maritime zones in the Southern Ocean

Maritime jurisdiction in the Southern Ocean raises several challenging legal issues, the most significant being the status of waters adjacent to the Antarctic continent. To a lesser extent the status of remote area EEZs in the Southern Ocean has been questioned. The geographical coverage of *CCAMLR* is a useful starting point for an examination of international, regional and national legal regimes applicable in the Southern Ocean.

5.2.1 *CCAMLR* Area of Application

The Convention applies to an expansive area of ocean. All living marine resources south of a physiological boundary referred to as the 'Antarctic Convergence' fall within its jurisdiction.[15] This is a larger area than that covered by the *Antarctic Treaty*[16] which applies only to the area south of 60 degrees Latitude.[17] Article I of *CCAMLR* states the Convention applies to:

> [...t]he Antarctic marine living resources of the area south of 60 degrees latitude and to the Antarctic marine living resources of the area between that latitude and the Antarctic Convergence which forms part of the Antarctic marine ecosystem.

The Antarctic Convergence is a naturally occurring boundary between the relative warmth of sub-Antarctic waters and the much colder, denser waters of the Southern Ocean. Essentially, cold less saline water from melting ice sheets and icebergs meets the relatively warmer salt water moving south and forms a natural divide between the Antarctic marine ecosystem and adjacent ecosystems. Scientists have identified this natural boundary as an appropriate means of applying an ecosystem approach to the management of Antarctic marine living resources, for it is within this mix of

---

with the delays by governments to take effective action against IUU fishing' CCAMLR-XXII (2003) Report of the Twenty Second Report of the Commission, paragraph 14.25. The fishing industry has also commented on *CCAMLR*'s role. See, L. Tucak, 'Sink poaching boats so they don't return – fishing boss' *The Australian* 19 February 2002, 5.

14 This argument is also put by the CCAMLR Executive Secretary. See, *The Mercury* 'Editorial – Letter to Editor by Dr D. Miller, Executive Secretary, *CCAMLR*' 13 November 2003. 14.

15 *CCAMLR,* Article 1(1) and (4).

16 Opened for signature 1 December 1959, 402 UNTS 71 (entered into force 23 June 1961). ('*Antarctic Treaty*'). As of September 2005 there were 44 signatories to the *Antarctic Treaty*, 27 States are Consultative Parties.

17 *Antarctic Treaty,* Article VI.

waters that conditions create a perfect haven for the beginnings of marine life. The Convergence also acts as a natural barrier to fish migration[18]

Recently the term Antarctic Polar Front has been used in preference to the Antarctic Convergence.[19] However the area of application remains unchanged. The Convention's boundary has been charted and at first glance it appears to wander haphazardly throughout the South Indian, South Atlantic and Southern Pacific Oceans. These waters encompass what has become known as the Southern Ocean, a vast body of water in the South Polar Region. The Southern Ocean has been estimated to extend over an area of seas approximately 37.5 million square kilometres[20] with large expanses of pack ice occurring closer to the Antarctic continent. Map 5.1 illustrates the Convention boundary.

Under international law, third States are not obliged to comply with *CCAMLR* conservation measures and may assert that the Convention Area is high seas. They are able to do this, to the detriment of those States which do comply with restrictions on fishing quotas, catch methods and season length, because of the *pacta tertii* rule.[21] In instances where third States allow their flagged vessels to engage in fishing within *CCAMLR* waters (or indeed any other RFMO area of application) contrary to *CCAMLR* conservation measures, such States have been described as 'free riders' by various authors.[22]

18 Arthur Watts, *International Law and the Antarctic Treaty System* (1992) 151.

19 The Antarctic Polar Front is defined as the zone where the colder less saline waters flowing north from the Antarctic meet the warmer more saline waters flowing from the south in the Atlantic, Indian and Pacific Oceans. The mean position is between 45 degrees and 60 degrees south. See, D. Miller, E. Sabourenkov and N. Slicer, 'Unregulated Fishing- The Toothfish Experience' 2003. (copy on file)

20 D. Rothwell and C. Joyner, 'The Polar Oceans and the Law of the Sea' in A. Oude Elferink and D Rothwell (eds) *The Law of the Sea and Polar Maritime Delimitation and Jurisdiction* (2001) 3.

21 This statement is not a criticism of international law rather a statement reflecting the reality of the international political circumstances whereby decisions about membership of RFMOs and compliance with RFMO measures is dependent upon State consent.

22 For example Hayashi states a growing number of fishing vessels are becoming 'free riders' by circumventing regional measures by not becoming parties to RFOs or by not reporting their catch properly. M. Hayashi, 'Illegal, Unreported and Unregulated (IUU) Fishing: Global and Regional Responses' in D. Caron and H. Scheiber (eds) *Bringing New Law to Ocean Waters* (2004) 95, 96.

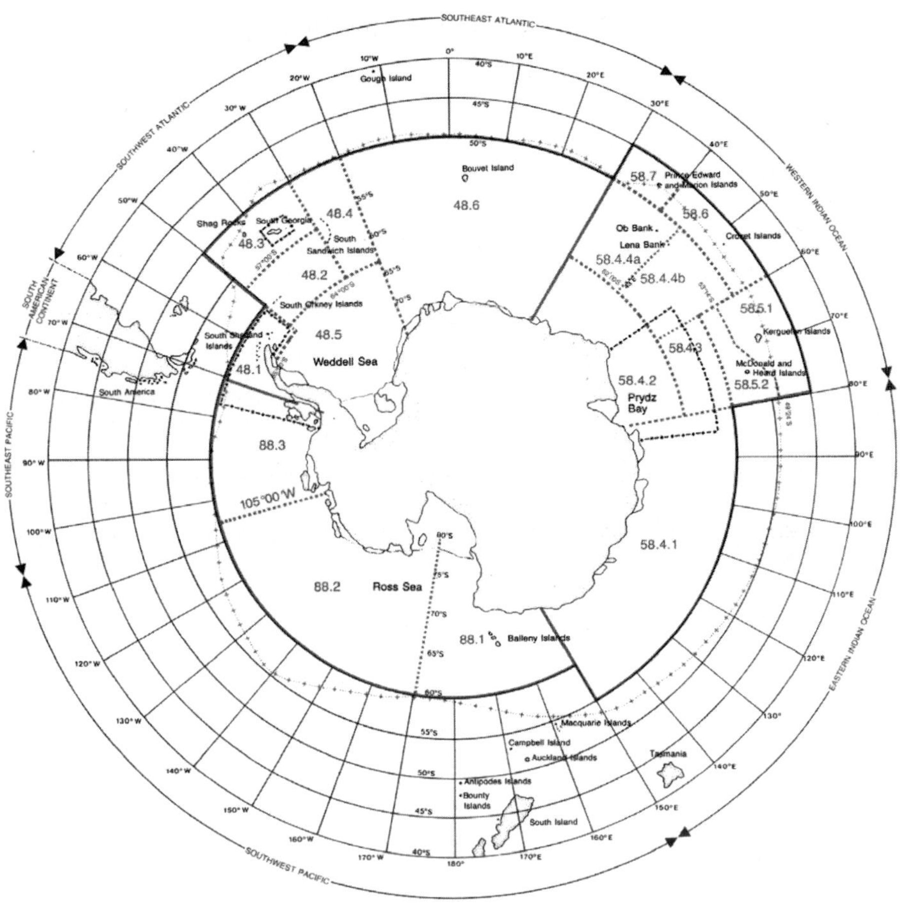

Map 5.1 CCAMLR Convention Area with location of sub-Antarctic Islands

As is evident on Map 5.1, the practical effect of extending the application of *CCAMLR* beyond 60 degrees south latitude is the inclusion of a number of sub-Antarctic islands within the *CCAMLR* area of application. Territorial sovereignty over some of these islands was expressly recognised at the close of the *CCAMLR* Conference in 1980 with the adoption of the Chairman's statement.[23] Beyond the areas of coastal State jurisdiction, the Southern Ocean is properly classified as high seas, albeit subject to a RFMO which creates obligations for State parties to *CCAMLR*.

23 Statement by Chairman at the conclusion of the Conference on the Conservation of Antarctic Marine Living Resources dated 19 May 1980 ('Chairman's statement'). Annexed to the text of *CCAMLR*, see paragraph 4. A copy of the Chairman's Statement is at Annex A of this Chapter.

5.2.2 Coastal State Claims in the Southern Ocean

The effect of the declared EEZs based on territorial sovereignty over a number of sub-Antarctic islands has been to displace high seas fishing rights and introduce coastal state jurisdiction to remote marine areas. Table 5.1 lists the sub-Antarctic islands subject to territorial claims. Their location can be found on Map 5.1 Starting from South Georgia (approximately 55'S 44'W) the Toothfish harvest has moved eastwards over ten to fourteen years. Toothfish populations in waters adjacent to the Prince Edward and Marion Islands (South African), Crozet and Kerguelen Isles (French) and the Heard and McDonald Islands (Australian) have all been affected by illegal fishing.

Country	Island Possession
Australia	Heard and McDonald Islands (sub-area 58.5.2) Macquarie Island
France	Crozet Island (sub-area 58.6) Kerguelen Isles (sub-area 58.5.1)
South Africa	Prince Edward Island and Marion Island (sub-area 58.7)
Norway	Bouvet Island (sub-area 48.6) Peter 1st Island
New Zealand	Balleny Islands (sub area 88.1) Scott Islands
United Kingdom/ Argentina[24]	Shag Rocks and South Georgia Island (sub area 48.3) South Sandwich Islands (sub area 48.4) Laurie Island, South Orkney Island, South Shetland Island.

Table 5.1 Sub-Antarctic island possessions

24 Argentina and the United Kingdom remain in dispute over sovereignty to these islands. The issue is still raised at *CCAMLR* meetings. For example at *CCAMLR*-XXI (2002) Argentina and the UK engaged in dialogue over the disputed islands. See paragraphs 11.18 -11.20 and 14.10. The disputed sovereignty became more of an issue for Argentina with the proposed centralised VMS which entails dual reporting to the *CCAMLR* Secretariat and flag State. The Secretariat would then distribute the VMS data to all *CCAMLR* contracting parties. The UK, as a *CCAMLR* contracting party, would then be privy to data detailing the position of Argentinean vessels which would be politically undesirable to the Argentinean government.

5.2.3 The Relevance of the Chairman's Statement

The adoption of the Chairman's statement was prompted by the firm position taken by the French government to reserve the right to exclude any *CCAMLR* conservation measure from applying to French declared EEZs within the Convention Area. The Statement recognises the sovereignty of France over its claimed sub-Antarctic islands in addition to the right to chose to enforce conservation measures, whether of national or *CCAMLR* origin.[25]

The French position has benefited other States with sub-Antarctic island territories because State sovereignty has been acknowledged in relation to waters adjacent to other islands within the Convention Area.[26] Australia's sovereign rights with respect to Heard and McDonald Islands, as they are undisputed by any *CCAMLR* Contracting Parties, are included within the scope of this recognition.

Some authors have made the comment that the Chairman's Statement, in which coastal State rights over sub-Antarctic Islands within the Convention Area are recognised, has no legal status. The argument is based first, upon the assertion that the Statement is redundant because it adds nothing to the international legal position,[27] and secondly, because it is of questionable legal status for it is neither part of the Convention text nor a reservation (by the French) to the Convention.[28]

In relation to the first argument, it is true that coastal State rights as recognised under the *LOSC*, are already re-affirmed under *CCAMLR*. Article IV 2(b) of *CCAMLR* states that 'the right or claim or basis of claim to exercise coastal State jurisdiction under international law within the area to which this Convention applies' is not affected by the Convention. This includes rights accorded under the *LOSC* over waters adjacent to sub-Antarctic islands of undisputed sovereignty. The position of States with respect to those sub-Antarctic islands over which sovereignty is disputed, is addressed in Article IV 2(c). The full text of Article IV is reproduced following.

---

25 Ibid.

26 Ibid. paragraph 5.

27 That is the right to declare an exclusive economic zone exists under Part V of the *LOSC* and by corollary, so does the right of the coastal State to choose not to be bound by *CCAMLR* Measures with the exclusive economic zone.

28 F.M. Auburn, *Antarctic Law and Politics* (1982) 222.

CCAMLR, Article IV

1. With respect to the Antarctic Treaty area, all Contracting parties, whether or not they are Parties to the Antarctic Treaty, are bound by Article IV and VI of the Antarctic Treaty in their relations with each other.
2. Nothing in this Convention and no acts or activities taking place while the present Convention is in force shall:
   a) constitute a basis for asserting, supporting or denying a claim to territorial sovereignty in the Antarctic Treaty area or create any rights of sovereignty in the Antarctic Treaty area;
   b) be interpreted as a renunciation or diminution by any Contracting Party of, or as prejudicing any right or claim or basis of claim to exercise coastal state jurisdiction under international law within the area to which this Convention applies;
   c) be interpreted as prejudicing the position of any Contracting Party as regards its recognition or non recognition of any such right, claim or basis of claim;
   d) affect the provision of Article IV, paragraph 2 of the Antarctic Treaty that no new claim, or enlargement of an existing claim, to territorial sovereignty in Antarctica shall be asserted while the Antarctic Treaty is in force.

Critics of the usefulness of the Chairman's Statement also argue that it is merely a reaffirmation of the right of any contracting party to make a reservation to any conservation measure adopted by the Commission.[29] However the effect of the Chairman's Statement for France, Australia or any other State with sub-Antarctic island possessions, is that the *CCAMLR* Commission may not proceed to adopt specific conservation measures in relation to waters adjacent to islands within the Convention Area over which the existence of State sovereignty is recognised by all Contracting parties, unless the relevant coastal State is in agreement.[30] This proposition differs from a reservation to a conservation measure. Further, it is recognised that States with sub-Antarctic island possessions have the right 'to

[29] *CCAMLR,* Article IX (6)

[30] Chairman's Statement, above, paragraphs 2 and 5.

promulgate any national measures which it might deem appropriate.'[31] Such national measures may impose stricter conservation controls than the *CCAMLR* measures and given the national sovereignty over the islands and adjacent EEZs, the national measures and controls take priority.

5.2.4 The Relevance of Article 121 *LOSC*

An argument in support of the Chairman's Statement is that the international recognition provided in the Statement is valuable, even if it merely lends more weight to the legitimacy of the remote area EEZs. This is more important than one might think, for as recently as December 2002, the right of coastal States to declare EEZs offshore from remote territorial possessions, was questioned, albeit by a single dissenting Judge of ITLOS.[32] Justice Vukas has expressed his reservation on more than one occasion and has been in dissent on this point on both occasions. In relation to the French Kerguelen isles he stated:

> In my view it is highly questionable whether the establishment of an exclusive economic zone off the shores of these uninhabitable and uninhabited islands (according to Captain Yves-Joseph de Kerguelen-Tremarec) is in accordance with the reasons which motivated the Third United Nations Conference on the Law of the Sea to create that specific legal regime and with the letter and spirit of the provisions on the exclusive economic zone contained in the United Nations Convention on the Law of the Sea.[33]

This statement is of limited value on a number of grounds. First, although not directly raised in the proceedings for prompt release before ITLOS,[34] no other Tribunal judge deemed it necessary to address the issue raised by Vukas in their judgement.[35] It is argued this is because it is simply not a legal issue on the facts, given the accepted sovereignty over several sub-Antarctic islands and particularly the French and Australian islands. This raises the second ground of objection to Vukas' statement namely whilst the issue he alludes to (that of rocks v. islands) is interesting, it is not relevant on the facts.

---

31 Ibid, paragraphs 2 and 3.

32 See the *Monte Confurco* Case (*Seychelles v France*) Case No .6 ITLOS Judgement dated 18 December 2000, Declaration of Judge Vukas, paragraph 3 and the *Volga* Case (*Russian Federation v Australia*) Case No. 11 ITLOS. Judgement dated 23 December 2003. Declaration of Judge Vukas, paragraph 2.

33 The *Monte Confurco Case*, Ibid.

34 Although raised indirectly it is still relevant for if there is no recognition of a right to declare an EEZ, then there is no coastal State right of boarding and arrest under Article 73, LOSC.

35 Judge Vukas has raised this point in two separate dissenting opinions in two separate cases. No other judge sitting on the *Monte Confurco* or *Volga* cases mentions his position. Even the other dissenting judges do not take up Vukas' point and it is reasonable to draw from the resounding silence a lack of support amongst the ITLOS bench for the views of Vukas on this point of law.

Under Article 121 of the *LOSC*, islands may sustain an EEZ, rocks may not. The specific wording of Article 121(3) states: 'rocks which cannot sustain human life or economic life on their own shall have no exclusive economic zone or continental shelf.'

Islands are broadly defined in Article 121(1) which states, 'an island is a naturally formed area of land surrounded by water, which is above water at high tide'. The ambiguity within Article 121 created by the reference to rocks without the provision of a specific definition was raised during UNCLOS III.[36] For the purposes of this Chapter, a close examination of the meaning of 'rock' or a debate on whether in fact the word 'island' would have been more appropriately used in Article 121(3),[37] is not required. Furthermore the territorial sovereignty over the Heard and McDonald Islands is not in dispute, nor is the right to declare an EEZ and continental shelf to the full limits allowed under international law.

It is however useful to note that in the *Jan Mayen* Case,[38] an island of similar size to Heard Island, the ICJ found the island was sufficient size to 'generate potential title to the maritime areas recognised by customary law'.[39] The ICJ said in response to Denmark's submission that the maritime boundary between Eastern Greenland and Jan Mayan should be drawn 200 miles from the base lines on the coast of Eastern Greenland, that:

> The coast of Jan Mayan, no less than that of Eastern Greenland, generates potential title up to a limit of 200 miles from its baselines. To attribute to Norway merely the residual area left after giving full effect to the eastern coast of Greenland would run wholly counter to the rights of Jan Mayan and also to the demands of equity.[40]

Jan Mayan has an area of 373 square kilometres. Heard Island has an area of 368 square kilometres. The question of whether Heard Island or Macdonald Island can sustain human habitation or support economic life of their own is complex

---

36 See B. Kwiatkowska and A. Soons, 'Entitlement to maritime areas of rocks which cannot sustain human habitation or economic life of their own' (1990) 21 *Netherlands Yearbook of International Law* 139, 140. In particular see the statement by Hogson and Smith in reference to Article 121(3) that 'it is essential that a rock be defined objectively so as to remove all doubts as to which rocks would be affected by this provision', 140 and n 3.

37 There are many papers written on this issue. See generally, Kwiatkowska and Soons, 'Entitlement to maritime areas of rocks' above, R. Churchill, 'Maritime Delimitation in the Jan Mayan Area' (1985) 9 *Marine Policy* 16, E.D. Brown, 'The Continental Shelf and the Exclusive Economic Zone. The Problem of Delimitation at UNCLOS III' (1977) 4 Maritime *Policy and Management* 390 and E.D. Brown, *Sea-bed Energy and Minerals: The International Legal Regime* Vol.1 The Continental Shelf (1992) 37-43.

38 *Case concerning Maritime Delimitation in the Area between Greenland and Jan Mayen (Denmark V Norway)* [1993] ICJ Rep. (the *Jan Mayen Case*) Jan Mayan is a Norwegian possession and East Greenland Danish. The issue of a maritime boundary between Eastern Greenland and Jan Mayan was in dispute.

39 Ibid, paragraph 70.

40 Ibid.

and difficult to answer with any certainty.[41] For example, scientific stations have been established in the past.[42] Further, there are a number of instances where small and remote islands have been accorded full effect when maritime boundaries are settled and it is argued that the sub-Antarctic islands are important examples of such islands.[43] In the French-Anglo Award the British argued that 'there is clear evidence that contemporary British practice treats the Eddystone Rocks as an island for all purposes, including the use of the low-water line around this island for the measurement of maritime zones.[44]

The third response to Vukas' statement rests upon the fact that whilst States have raised reservations to the declaration of an EEZ offshore to the Australian Antarctic Territory, no State has objected to the EEZ declared offshore to the Heard and McDonald Islands nor to Australia's exertion of sovereignty over the islands.

In 1994 Australia declared a 200 mile EEZ in relation to her external territories, including the Australian Antarctic Territory.[45] Whilst the Australian position is that that the maritime zone is simply a reflection of Australia's territorial sovereignty in Antarctica, it is significant that the official government policy is to limit the application of domestic fisheries laws within this zone to Australian nationals and vessels.[46] The USA in particular has objected to any claims over waters adjacent to the Antarctic continent, asserting that they remain high seas and are open to all mankind.[47]

Australia gained sovereignty over the Heard and McDonald Islands in 1953 and a fishing zone was declared 1979.[48] In the *Anglo Norwegian fisheries* case the ICJ held that the Norwegian practice of drawing straight baselines to delimit its territorial sea had been tolerated by foreign States for over 60 years.[49] The ICJ stated:

---

41 See generally Kwiatkowska and Soons, 'Entitlement to maritime areas of rocks' above.

42 See Churchill, 'Maritime Delimitation in the Jan Mayan Area' above n 37. Details on Heard Island are provided in Chapter 7 and at <www.deh.goc.au/heritage/worldheritage/sites/anatrct/index>. See also E.D Brown, *Sea-Bed Energy and Minerals*, above, 38-39. At the time of writing Brown argued that this was a difficult test to apply.

43 Indeed Churchill, in 'Maritime Delimitation in the Jan Mayan Area' above, cites a number of remote unpopulated islands including Jan, Mayan, Clipperton, Crozet, Kerguelen and Clarion.

44 *Jan Mayan Case* [1993] above ,paragraph 128.

45 Proclamation dated 26 July 1994.

46 See n66 below.

47 D.Rothwell and C.Joyner, 'The Polar Oceans and the law of the Sea' abov, 2. See also, C. Joyner, 'The Exclusive Economic Zone and Antarctic' (1981) 21(4) *Virginia Journal of International Law* 691 at 709 and C. Joyner, 'The Antarctic Treaty System and the Law of the Sea' (1995) 10(2) *International Journal of Marine and Coastal Law* 301 at 303. This position was reiterated at the Southern Ocean Workshop, Sydney University, 19 October 2001 in a presentation by C. Joyner.

48 See Chapter 7 for detail.

49 *Anglo-Norwegian Fisheries* Case (*UK v. Norway*) [1951] ICJ Rep. 116.

> The notoriety of the facts, the general toleration of the international community, Great Britain's position in the North Sea, her own interest in the question, and her prolonged abstention would in any case warrant Norway's enforcement of her system against the United Kingdom.[50]

It is argued by analogy that none of the 24 *CCAMLR* Commission members or eight acceding States have raised any objection to Australian sovereignty over the Islands or the right to declare an EEZ which arises from the fact of territorial sovereignty. It is noted that one IUU fisher arrested within the Heard and McDonald Islands' fishing zone is quoted as stating: 'We…the fishermen…do not understand the Australian waters as Australian waters'[51] perhaps meaning that Australian sovereignty was not recognised. However, there has never been any formal challenge to Australia's right to declare a 200 mile EEZ around the Heard and McDonald Islands. In any event, that particular defendant fisherman pleaded guilty to the charge of fishing illegally under sections 100(2) and 100A of the *Fisheries Management Act.*[52]

For the reasons outlined above it is concluded that fishing within the Australian declared EEZ located in the Southern Ocean (and indeed other declared EEZs by other States where the territorial sovereignty is not in dispute) is properly classified as illegal within the framework of IUU fishing.

5.2.5 The Status of *CCAMLR* Seas

The second legal layer, which applies not just to fish stocks within the Southern Ocean, but to all living marine resources, is the management regime established under the auspices of *CCAMLR*. The stated objective of the Convention is the conservation of Antarctic marine living resources.[53] Non-compliance with conservation measures adopted by the Commission under the Convention, by both contracting parties (Commission members and acceding States) and third States has been problematic. This is partly because non-contracting parties have not consented to the authority of the Commission to regulate the Southern Ocean, and continue to harvest fish stocks in the exercise of their high seas rights. Nevertheless for contracting parties the conservation measures adopted by the *CCAMLR* Commission represent binding regional measures.

With reference to the definition of IUU fishing, activities may be classified as both illegal and unreported (for contracting parties) or unregulated, but legal

---

50 Ibid.

51 ABC, 'The Toothfish Pirates' *4 Corners,* 30 September 2002. <www.abc.net.au/4corners/archives/ 2002b_Monday 30 September2002ht> (14 October 2002). The fisherman (Sanchez) was interviewed for the program.

52 See Table 7.1 Chapter 7 for details on sections 100(2) and 100A of the *Fisheries Management Act* 1991 (Cth).

53 *CCAMLR* , Article II(1)

(for non-contracting parties). Therefore, there will be differences in the approach taken by RFMOs to tackle the problem of IUU fishing depending upon the legal context in which it is performed. However, for the purpose of this study which examines the implementation of measures to deter IUU fishing by *CCAMLR* and Australia in the Southern Ocean in the absence of effective flag State control; the measures identified in this Part and Part 5 are applicable to all forms of IUU fishing.

The final legal regime to be factored into any analysis of the Southern Ocean is the wider application of international fisheries law, particularly the right to fish the high seas. *CCAMLR* expressly acknowledges the rights of States on the high seas and more importantly the understanding that those high seas rights remain unaffected. This is achieved through the adoption of a number of Articles from the *Antarctic Treaty*.[54] In particular, Article IV of *CCAMLR* states that all *CCAMLR* Contracting Parties are bound by Articles IV and VI of the *Antarctic Treaty*. Article VI of the *Antarctic Treaty* reserves high seas fishing rights within the Treaty area. It has been reproduced below.

Antarctic Treaty, Article VI

The provisions of the present treaty shall apply to the area south of 60 degrees South Latitude, including all ice shelves, but nothing in the present Treaty shall prejudice or in any way affect the rights or the exercise of the rights of any State under international law with regard to the high seas within the area.

It is noted that the reservation of high seas rights is extended to 'any State'. Thus, *Antarctic Treaty* consultative parties and *CCAMLR* contracting parties (through the adoption of Article VI) retain high seas rights within the *Antarctic Treaty* area in addition to non-contracting parties. The difference in geographical area of application between *CCAMLR* and the *Antarctic Treaty* is slight given that the former Convention applies generally to the area south of 60 degrees south latitude and north of this latitude where the Antarctic Convergence tracks further north.[55]

54 Opened for signature 1 December 1969, 402 UNTS 71 (entered into force 23 June 1961). Articles 1 and V of the *Antarctic Treaty* are adopted by Article III *CCAMLR*. Articles IV and VI of the *Antarctic Treaty* are adopted by Article IV of *CCAMLR*.

55 *CCAMLR*, Article 1 and 4,

As discussed, there has been debate about the status of some marine areas adjacent to the *Antarctic Territory*. Since the adoption of the *Antarctic Treaty* in 1959 those States who were not amongst the original territorial claimants in the Antarctic, have maintained the position that the high seas extend from the low water mark along the Antarctic coastline.[56] The difficulty for Antarctic claimant States is that any claim to Antarctic maritime zones under the *LOSC* is based on territorial sovereignty. The principle extends back to the initial exercise of authority by maritime States over the belt of waters adjacent to their coastlines. It was recognised that the extension of sovereignty over territorial seas was dependent on the existence of territorial sovereignty. The 1958 *Convention on the Territorial Sea and Contiguous Zone* confirmed the principle.[57] Under Article 55 of the *LOSC*, an EEZ is defined as an area beyond and adjacent to the territorial sea of the coastal State. In turn, a territorial sea is defined (in almost identical words as the *Convention on the Territorial Sea and Contiguous Zone*) as a belt of sea adjacent to the land territory and its internal waters, to a breadth of 12 nautical miles.[58]

Two arguments arise in opposition to the various claimant State declarations of maritime zones, purportedly based on sovereignty to Antarctic territory and made pursuant to the *LOSC*.[59] First, the territorial claims to Antarctica have not been recognised at international law.[60] In fact, they have been specifically 'frozen' under Article IV of the *Antarctic Treaty* which effectively sidesteps the need to determine the validity of such claims. Second, the declaration of an EEZ or territorial sea is arguably in breach of Article IV which prohibits any new claim to territorial sovereignty, or any enlargement of an existing claim, from being made.[61] Article VI of the *Antarctic Treaty*, which has been adopted in Article IV of *CCAMLR* is reproduced on the following page.

Australia declared an EEZ with respect to all her external territories, including Antarctica, in 1994.[62] This claim has not been expressly rejected or accepted by any

56 Auburn, *Antarctic Law and Politics* above, 221. There is no consensus on where the low water mark starts because of the shifting ice shelf, however this may not present as a practical issue due to the 'freezing' of sovereignty under the *Antarctic Treaty* and the fact that no States have sought to enforce their law of the sea rights with respect to their claimed portion of the Antarctic.

57 Opened for signature 29 April 1968, 516 UNTS 205 (entered into force 10 September 1964). Article 1 of the *Convention on the Territorial Sea and Contiguous Zone* states: 'The sovereignty of a State extends beyond its land territory and its internal waters, to a belt of sea adjacent to its coast, described as the territorial sea.'

58 *LOSC*, Articles 2 (1) and 3.

59 In addition to Australia, the United Kingdom established a 200 mile EEZ around the disputed South Georgia and South Sandwich Islands in 1993. Chile claimed in 1991 a *mar presencial* over 20 million square kilometres of sea extending west to Easter Island and south to encompass waters adjacent to the Chilean sector claim on the Antarctic continent.

60 Australia's claim to 42% of the continent has been recognised by just four States: New Zealand, France, Norway and Britain – all Antarctic claimant States themselves. Non-claimant Consultative Parties including Belgium, Japan, South Africa, the then Soviet Union and USA refuse to recognise territorial claims. See Joyner, 'The Exclusive Economic Zone in Antarctica' above, 703.

61 *Antarctic Treaty*, Article IV (2).

nation. However, the USA has continued to maintain the position that in the absence of any recognised coastal sovereign, the regime of high seas actually extends up to the Antarctic coastline.[63]

The Australian position rests upon the argument that the rights afforded to coastal States under the *LOSC* exist as a matter of international law and the formal declaration of a maritime zone under domestic law is not, therefore, an enlargement of an existing jurisdiction. This argument has found support amongst some Australian authors.[64] In support of the Proclamation of the EEZ, Australia has been reported as stating:

> The proclamation of an EEZ adjacent to the Australian Antarctic Territory is merely a reflection of Australia's territorial sovereignty in Antarctica and of the progressive development of international law through the law of the sea.[65]

This point of disagreement over where the high seas commence has not been tested and is unlikely to be. This is because as a matter of policy, the Australian Government has not sought to enforce any coastal State rights within the declared Australian Antarctic EEZ.[66] Indeed, this zone has been specifically exempted from the application of the *Fisheries Management Act 1991*.[67] In stark contrast, Australia has vigorously and consistently defended her rights within the Heard and McDonald Islands' EEZ.

The weight of current international practice, including that of Australia in choosing not to exert rights with respect to foreign fishers within the Australian Antarctic EEZ, suggests that the better conclusion is that the legal status of the seas surrounding the Antarctic continent is that of high seas.[68]

---

62 Proclamation dated 26 July 1994.

63 Rothwell and Joyner, 'The Polar Oceans and the Law of the Sea' above, 2. See also Joyner, 'The Exclusive Economic Zone and Antarctica' above.

64 See for example Gillian Triggs, *The Antarctic Treaty Regime – Law, Environment and Resources* (1987) 103.

65 T. Scovazzi, 'The Antarctic Treaty System and the New Law of the Sea: Selected Questions' in Francioni and Scovazzi (eds) *International Law for Antarctica* (1996) 382.

66 Ibid. See also the judgment in *Humane Society International Inc. v. Kyodo Senpaku Kaisha Ltd* [2005] FCA678 paragraphs 13-14. In this case concerning a claim by the Humane Society, the Commonwealth Attorney-General stated in relation to the application of Australian laws to nationals of other States that, "...the Government of Japan would regard any attempts by Australia to enforce Australian Law against Japanese vessels and its nationals in the Antarctic EEZ to be a breach of international law on Australia's part...Australia...has not enforced the laws of the Commonwealth in Antarctica against nationals of other States which are parties to the Antarctic Treaty."

67 The definition of the Australian Fishing Zone within the *Fisheries Management Act 1991* excludes waters that are exempted by Proclamation under section 11.

68 Although some authors have recently commented that this issue remains unresolved. A. O. Elferink and D. Rothwell, 'Challenges for Polar Maritime Delimitation and Jurisdiction: The Current Regime and its Prospects' in A. O. Elferink and D.Rothwell (eds) *The Law of the Sea and Polar Maritime Delimitation* (2001) 350.

One can conclude from the above examination that the waters surrounding the Antarctic continent are high seas as defined in Article 86 of the *LOSC*. Thus the *CCAMLR* management regime applicable to the seas surrounding the Antarctic continent does not, for non-members, alter the status of the seas as high seas. This legal distinction has been *CCAMLR*'s Achilles heel for the past twenty years. As a matter of practice many non-contracting party States to the *Antarctic Treaty* and *CCAMLR* continue to argue that neither Convention alters their right to assert high seas freedoms within the Southern Ocean. This is subject, of course, to those pockets of coastal State waters such as the Heard and McDonald Islands' EEZ. By way of example, Latvia, Portugal, Belize, Vanuatu and Panama fish within the Convention Area and do not comply with restrictions on catch limits, methods of harvesting or reporting requirements established via *CCAMLR* conservation measures.[69]

5.3 Formulating and Adopting conservation measures

CCAMLR was drafted with the primary objective of conserving Antarctic marine living resources.[70] Conservation is defined to include rational use.[71] The Convention was unique in its time and, for a fisheries convention, remains unique, because it has employed an ecosystem approach in what is essentially a fisheries management convention.[72] Existing regional fisheries conventions have adopted a more specific approach, applying to single fish species or fisheries resources within a particular area.[73]

Before commencing negotiations for *CCAMLR*, the *Antarctic Treaty* consultative parties expressed initial concerns about the developing krill fishery, and were mindful of krill's pivotal role in the Antarctic marine ecosystem.[74] In this context, Article II(3) of *CCAMLR* requires any harvesting and associated activities to be

[69] *CCAMLR*-XV (1996) Report of the Standing Committee on Observation and Inspection, paragraphs 1.53-1.58. *CCAMLR*-XVI (1997) Report of the Standing Committee on Observation and Inspection, paragraphs 1.7-1.26.

[70] *CCAMLR*, Article II (1).

[71] *CCAMLR,* Article II (2). This term implies that the Convention is not a pure conservation Convention for rational use contemplates harvesting of the marine living resources.

[72] *CCAMLR* , Article I(2) defines Antarctic marine living resources to mean the populations of fin fish, molluscs, crustaceans and all other species of living organisms, including birds, found south of the Antarctic Convergence. Article VI excludes application to whales and seals. Whilst some Conservations Measures do address the impact on sea birds, in practice the main work of the Commission has been to manage and set quotas for particular fish stocks.

[73] For example the *Convention on Future Multi-Lateral Cooperation in the Northwest Atlantic Fisheries* opened for signature 24 October 1978, 1135 UNTS 367 (entered into force 1 January 1979) applies to fisheries resources of the Convention Area, Article II. The *Convention on the Conservation and Management of Pollock resources of the Central Bering Sea* opened for signature 11 February 1994 (entered into force for parties on varying dates) applies to Pollock resources within the Convention Area. Article II. The *Convention for the Conservation of Southern Bluefin Tuna* (1994) ATS 16, applies to Tuna.

[74] Kock, ' Fishing and Conservation in the Southern Ocean' above, 5-6.

conducted in a sustainable manner with due regard to ecological relationships between harvested and dependant species.

It is this emphasis on the integrity of the marine ecosystem, in preference to individual species, which sets the Convention apart from other regional fishing conventions. It has meant that decisions on catch levels, fishing methods and fishing seasons are to be made with reference to available knowledge on the marine ecosystem and not just the target species. In addition, measures are to be formulated on the basis of the best scientific evidence available. These constraints on the decision making process have on occasions made the tasks bestowed upon the Scientific Committee and the Commission more difficult.[75]

The Convention addresses the issue of management through the formulation and adoption of conservation measures which may cover a wide range of matters including: the quantity of any species which may be harvested, regulated by species and regions or sub-regions; protected species; the size, age and other biological limits applying to particular species to be harvested; season start and close dates; specific closed seasons or regions; or methods of harvesting- including fishing gear to be used.[76] The adoption of such other measures as required provides further flexibility to the *CCAMLR* process. Article IX (2)(i) states conservation measures may include:

> the taking of such other conservation measures as the Commission considers necessary for the fulfilment of the objective of this Convention, including measures concerning the effects of harvesting and associated activities on components of the marine ecosystem other than the harvested populations.[77]

The *CCAMLR* Commission adopts measures during its annual meetings, with reference to the recommendations and advice of the Scientific Committee.[78] Decisions on matters of substance, which presumably includes all conservation measures, are taken by consensus.[79] This means a single vote by one of the Commission members [80] can block measures or delay their adoption until substantially amended.

75 For example, during the 2003 meeting, evidence of the Toothfish population in the Indian Ocean constituting a meta-population was discussed. See, *CCAMLR*-XXII (2003) Report of the Twenty Second Meting of the Commission, paragraphs 11.2-11.3. Several member States made the comment that further work was required before such a conclusion could be drawn. The consequences of reaching an agreement that the Toothfish in the Indian Ocean is a meta-population are significant in terms of the management of the Toothfish as a transboundary stock.

76 *CCAMLR*, Article IX (2).

77 Ibid.

78 *CCAMLR*, Article IX(4).

79 *CCAMLR*, Article XII(1).

80 Under Article VIII membership of the Commission includes contracting parties which participated in the first meeting at which the Convention was adopted and acceding States during such time as the acceding State is engaged in research or harvesting activities in relation to the marine resources to which the Convention applies.

In practice the consensus requirement has delayed the adoption of conservation measures;[81] however, it has also worked to enhance the credibility of the Convention. This is because measures are only adopted when all Commission members are in agreement and hence, more likely to throw their support behind compliance. One potential weakness within the Convention framework enables States to legally avoid compliance with conservation measures. *CCAMLR* Commission members have the right to notify the Commission of any intention not to comply with an adopted conservation measures within 90 days of the official receipt of notification of such measure.[82] However, in the 24 years since *CCAMLR*'s entry into force, there have been no recorded instances of notification after a conservation measure has been adopted.

Under Article IX(1)(f), conservation measures are to be formulated 'on the basis of the best scientific evidence available' which means that Commission is to refer to the reports of the Scientific Committee. In practice, the Scientific Committee has been restricted in its ability to conduct accurate stock assessments because of uncertainty about population levels, the level of the IUU catch and a lack of agreement on the range of populations.[83]

Finally, on an administrative note, at the 2001 Commission meeting, the *CCAMLR* Secretariat was requested to review the conservation measures numbering system.[84] In 2002 the Secretariat's review was adopted and existing measures were renumbered. Categories of measures were established, with each category being designated by a two-digit number. All compliance conservation measures are prefixed by number '10' Hence all of the conservation measures in Table 5.4, which lists initiatives to secure compliance with conservation measures in relation to fishing for the Toothfish, commence with the numeral '10'

5.4 Status of *CCAMLR* as a Regional fisheries management organisation

The role of RFMOs in promoting the sustainable management of high seas fisheries is expressly acknowledged in the *LOSC*, *Compliance Agreement*, *UN Fish Stocks Agreement* and the *IPO-IUU*.[85] The common theme in these international instruments is cooperation on a regional or sub-regional basis.

[81] A recent example is the implementation of the proposed 'dual reporting' for the Vessel Monitoring System which involves reporting to both the Secretariat and flag State. There was insufficient support for implementing a centralised System at the 2003 Commission meeting, however members did agree to a pilot dual reporting system open to all members wishing to participate. *CCAMLR*-XXII (2003) Report of the Twenty Second Meeting of the Commission, paragraph 10.12.

[82] *CCAMLR*, Article IX(6).

[83] CCAMLR-XXIII (2003) Report of the Twenty Second Meeting of the Commission, paragraphs 11.2-11.3.

[84] *CCAMLR*-XX (2001) Report of the Twentieth Meeting of the Commission, paragraph 9.4.

[85] See generally Chapter 4.

The *IPOA-IUU* addresses cooperation between States in the following terms:

> States should coordinate their activities and cooperate directly, and as appropriate through relevant regional fisheries management organisations, in preventing, deterring and eliminating IUU fishing.[86]

The *UN Fish Stocks Agreement* which has the specific objective of ensuring 'the long term conservation and sustainable use of straddling fish stocks and highly migratory fish stocks' places greater emphasis on regional organisations or arrangements as a vehicle for achieving sustainable fisheries management. Article 8(1) states that 'coastal States and States fishing on the high seas shall, in accordance with the Convention [the LOSC], pursue cooperation ...either directly or thorough appropriate sub-regional or regional fisheries management organisations or arrangements.' Further, where a RFMO (or arrangement) with the competence to establish conservation and management measures for particular straddling fish stocks exists, States shall give effect to their duty to cooperate by becoming members of such an organisation or by agreeing to apply the conservation and management measures established by such an organisation or arrangement.[87] Although the term 'organisation' is not defined in the Agreement, 'arrangement' is defined as:

> [A] cooperative mechanism established in accordance with the Convention and this Agreement by two or more States for the purpose, inter alia, of establishing conservation and management measures in a subregion or region for one or more straddling fish stocks or highly migratory fish stocks.

Conservation and management measures are defined as:

> [M]easures to conserve and manage one or more species of living marine resources that are adopted and applied consistent with the relevant rules of international law as reflected in the Convention and this Agreement [88]

This study examines regional and national responses within the Southern Ocean to IUU fishing and does so within the context of the global problems caused by IUU fishing. It is therefore concerned with measures or strategies to address the management of straddling fish stocks or highly migratory fish stocks (many of which are targeted by IUU fishers) only to the extent that the measures reviewed in Parts 4 and 5 are of broad application and applicable to all fish stocks targeted by IUU fishers. As such, the status of *CCAMLR* as a RFMO in terms of the *UN Fish Stocks Agreement* does not require close examination. However the emphasis on RFMOs as the vehicle for achieving sustainable fisheries management and a tool

86 *IPOA-IUU*, paragraph 28.

87 *UN Fish Stocks Agreement*, Article 8(3).

88 *UN Fish Stocks Agreement*, Article 1(1).

for combating IUU fishing is relevant and the status of *CCAMLR* as such a vehicle is reviewed below.

CCAMLR is, strictly speaking, a conservation organisation with a focus on the Antarctic ecosystem as distinct from a specialist fisheries organisation. Nevertheless *CCAMLR* is playing a crucial role in the management of the Patagonian Toothfish and the identification and implementation of conservation measures aimed at deterring IUU fishing. On this basis, the balance of this section examines the status of *CCAMLR* as a RFMO and concludes that the role of *CCAMLR,* not only as the sole RFMO in the Southern Ocean, but also as a RFMO which is leading international practice in the campaign against IUU fishing, is vital notwithstanding the semantics at play.[89]

To commence, the provisions of *CCAMLR* arguably satisfy the definitions within the *UN Fish Stocks Agreement* of an 'arrangement' and of 'conservation and management measures.' The Convention's 24 year history demonstrates the Commission's competence to establish conservation and management measures. Indeed one of the Commission's specified functions is to 'formulate, adopt and revise conservation measures on the basis of the best scientific evidence available.'[90]

Support for *CCAMLR* assuming an enhanced role as a RFMO, whether under the auspices of the *UN Fish Stocks Agreement* or otherwise, can be found in the reports of the Commission's annual November meetings.[91] The *CCAMLR* Commission has on several occasions noted the existence and subsequent entry into force of the *UN Fish Stocks Agreement.* At the 2002 Commission meeting, members agreed that the:

> Commission's role as a conservation organisation with responsibility for managing fisheries in the Southern Ocean gave it the attributes of a regional fisheries management organisation within the context of the United Nations. [92]

---

89 Indeed it would be inconsistent with the intent of Article 21(1) of the *UN Fish Stocks Agreement* to argue that *CCAMLR* is not a RFMO on the basis that it is a conservation organisation; or that it has not formally been declared a UN Fish Stocks RFMO (there is not reason for stating a declaration is required) or that only 12 *CCAMLR* contracting parties (from 31 members and acceding States) have ratified the UN Fish Stocks Agreement.

90 *CCAMLR* , Article IX(1)(f).

91 For example, see *CCAMLR*-XIII(1994) Report of the Thirteenth Meeting of the Commission, paragraph 12.20; *CCAMLR*-XIV (1995) Report of the Fourteenth Meeting of the Commission, paragraph 5.21; *CCAMLR* XV (1996) Report of the Fifteenth Meeting of the Commission, paragraphs 7.32-7.34; *CCAMLR*-XVI (1997) Report of the Sixteenth Meeting of the Commission, paragraph 5.25; *CCAMLR*-XVII (1998) Report of the Seventeenth Meeting of the Commission, paragraphs 5.8 and 5.20; *CCAMLR*-XIX (2000) Report of the Nineteenth Meeting of the Commission, paragraph 2.4; *CCAMLR*- XXI (2002) Report of the Twenty First Meeting of the Commission, paragraph 8.80 and 15.12 and *CCAMLR*-XXII (2003) Report of the Twenty Second Meeting of the Commission, paragraphs 18.5-18.7.

92 *CCAMLR* XXI (2002) Report of the Twenty First Meeting of the Commission, paragraph 15.2.

The Commission has also noted the competence of *CCAMLR* as a RFMO in relation to the management of the Patagonian Toothfish.[93] In 1996 Australia submitted a paper at the Commission's annual meeting entitled: '*The Relevance to CCAMLR of the UN Agreement relating to the Conservation and Management of Straddling Fish Stocks and Highly Migratory Fish Stocks.*'[94] In the paper, Australia recommended further examination by member States of the relationship between the two instruments and urged members to ratify the *UN Fish Stocks Agreement*. This sentiment was echoed by Norway at the 2003 meeting, where it was noted that very few members had ratified the *UN Fish Stocks Agreement*.[95]

The argument has been put forward that the objective of *CCAMLR*, namely the conservation of marine living resources, characterises the Convention as an ecosystem Convention unique to the Southern Ocean.[96] The significance of such a proposition is that one might then propose that *CCAMLR* is not a suitable vehicle for assuming the role of a competent RFMO under the auspices of the *UN Fish Stocks Agreement*. Another objection to *CCAMLR* assuming a formal role as the RFMO for the Southern Ocean is based on the fact that the Convention was negotiated under the *Antarctic Treaty* System and is therefore part of the management framework for the Antarctic. Yet another argument is that *CCAMLR* is not an FAO sponsored RFMO.

These arguments are not sustainable. First, the legitimacy of the regime established by *CCAMLR* is not diminished because decisions on conservation and management are based upon a holistic approach with regard to factors such as dependent species and by catch. In practice, the Commission adopts conservation measures addressing a range of matters which are within the usual scope of authority of a competent RFMO. Article IX states that Conservation measures may stipulate, *inter alia,* the quantity of any species to be harvested, the regions or sub-regions in which the harvest may take place, the length of any fishing season, the size, age and sex of any species to be harvested and the method of harvesting to be employed.[97]

In 2003 in a letter to Tasmanian Newspaper, *The Mercury*, the Executive Secretary of *CCAMLR* raised a question relevant to this issue. Responding to media criticism on the role and function of *CCAMLR*, Dr Miller asked a telling question which highlighted the competency of *CCAMLR* in managing Antarctic marine living resources (including the fish). He asked: 'What would the world do in the absence of *CCAMLR*?'[98] The answer is not worth pondering and the question illustrates the

---

93 Ibid.

94 *CCAMLR*-XV (1996) Report of the Fifteenth Meeting of the Commission, Document-*CCAMLR*-XV/12 Rev.1

95 *CCAMLR*-XXII (2003) Report of the Twenty Second Meeting of the Commission, paragraph 18.5.

96 Kock, 'Fishing and Conservation in the southern Ocean' above, 11. The statement was made that *CCAMLR* is an ecosystem convention.

97 *CCAMLR*, Article IX (2)(a) –(i).

98 Miller, *The Mercury*, 'Editorial- Letter to the Editor' above.

fundamental role the *CCAMLR* Commission plays in managing Antarctic marine living resources.

In relation to the proposition that *CCAMLR*'s position within the *Antarctic Treaty* System prevents it from assuming a role as the Southern Ocean RFMO, one might consider the *CCAMLR* membership. Whilst the overwhelming majority of *CCAMLR* members are also members of the *Antarctic Treaty*, there are two States which have joined *CCAMLR* because of its competence as a fisheries organisation in the Southern Ocean.[99] Namibia joined in 2001 because of its interest in fishing in the Indian Ocean and desire for support in managing its fisheries.[100] Additionally, several non-contracting parties of *CCAMLR* have indicated they will comply with various conservation measures to support the role of *CCAMLR* in managing the fishery for the Patagonian Toothfish.[101]

Although the *CCAMLR* Commission may not formally declare *CCAMLR* to be a RFMO under the auspices of the *UN Fish Stocks Agreement*, many factors support its position as a *de facto* fisheries organisation. The current Executive Secretary of *CCAMLR*, Dr Denzil Miller, is the head of the FAO RFMO Body, notwithstanding the fact that *CCAMLR* is not an FAO organisation.[102] In addition, *CCAMLR* was included in a 1999 FAO review of regional fisheries bodies.[103] Further support for *CCAMLR*'s role as a competent RFMO can be found in an examination of developments outside the *CCAMLR* Convention Area.

The *Convention for the Conservation and Management of Fishery Resources of the South East Atlantic Ocean* (SEAFC) is a new regional fisheries management regime negotiated under the *UN Fish Stocks Agreement*.[104] The primary objective of *SEAFC* is to ensure the long term conservation and sustainable use of the fishery resources in the *SEAFC* Convention Area.[105] The Convention area corresponds

---

[99] Vanuatu and Namibia are *CCAMLR* contracting parties but not parties to the *Antarctic Treaty*. All remaining 29 *CCAMLR* members are parties to the *Antarctic Treaty*.

[100] Personal communication with Dr Denzil Miller, Executive Secretary, *CCAMLR*, 14 November 2003.

[101] See for example the 2003 Commission meeting at which the Seychelles indicated that it was no longer involved in the Toothfish fishing industry. Its registration and ports would be closed to any fishing vessels engaged in IUU fishing. *CCAMLR*-XXII (2003) Report of the Twenty Second Meeting of the Commission, paragraph 8.9. China, another non-member, informed the Commission that, in support of the Catch Documentation Scheme, it had authorised the China Fisheries Association to sign certificates for any Toothfish unloaded in Chinese ports. *CCAMLR*-XXII (2003), paragraph 7.4.

[102] Personal communication with Miller, above.

[103] G. Lugten, 'A Review of Measures taken by Regional Fisheries Bodies to address Contemporary Issues' FAO *Fisheries Circular* No. 940 (1999), 22.

[104] Opened for signature 20 April 2001 (entered into force on 13 April 2003) Members include Namibia, Angola, South Africa, the United Kingdom (for island dependencies) the EC, Japan, Norway, the Russian Federation, the USA, Korea, Iceland, Poland and the Ukraine. SEAFC succeeds the 1969 International Commission for the Conservation of the Living Resources of the South East Atlantic. See, D.G. Miller, 'Management and Governance- Conventions and Protocols – SEAFC, WCPFC and SADC' (2003) 2-4. (copy on file)

[105] *SEAFC*, Article 2.

roughly to FAO statistical area 47, which is located north of area 48.6.[106] As such, *SEAFC* shares a common boundary, along 50 degrees south latitude, with *CCAMLR*. Although *SEAFC* has only recently entered into force, it is envisaged that regular communication on the management of transboundary stocks[107] will be required.[108] This interaction supports the role *CCAMLR* has assumed as a competent RFMO within the Southern Ocean.

In a similar vein, the *CCAMLR* Commission considered a proposal to extend its authority north of sub-areas 58.5, 58.6 and 58.7, to statistical areas 51 and 57. The proposal was based upon evidence that stocks of Toothfish in the south west Indian Ocean straddle *CCAMLR*'s northern boundary. [109] Significant catches have been reported from area 51 which have the potential to impact on management of the Toothfish population within the *CCAMLR* Convention Area.[110]

No competent RFMO currently exists in the high seas area within area 51. In light of this fact, Australia, at the 2002 Commission meeting, suggested that Article 1 of the Convention be amended to extend *CCAMLR*'s area of application north.[111] Whilst many States responded by stating that to amend the Convention in such a way would 'be a lengthy process' and 'may also impact on other' regional fisheries management organisations',[112] there was general support for the need to cooperate in the management of transboundary stocks. In this context, preliminary negotiations for the South Indian Ocean Fisheries Commission were noted, however 'in terms of timing, the establishment of this organisation was still uncertain.'[113]

The *CCAMLR* Secretariat circulated a possible solution to the absence of a competent RFMO in the waters north of the *CCAMLR* area of application during the 2003 Commission meeting.[114] The proposal contemplated the formation of a contingency zone in sub-area 51, with members supporting the management of

---

106 See Map 5.1 above in this Chapter.

107 Similar to straddling stocks, although not included within Article 63 of the *LOSC*, transboundary stocks straddle an area of high seas managed by two or more regional fisheries management organisations necessitating cooperation between the organisations on the management of the stocks.

108 Personal communication with Miller, above.

109 Ibid, see also, *CCAMLR*-XXII (2003) Report of the Twenty Second Meeting of the Commission, paragraphs 11.2-11.3. The Commission noted Scientific Committee reports that the Patagonian Toothfish stocks in the Indian Ocean are likely to be part of a meta-population with exchange of individuals between shelf areas across the Indian Ocean from east to west.

110 For example, in relation to FAO statistical area 51, 3.543 tonnes were reported for the 2002/2003 fishing year which has decreased from a reported catch of 10, 620 for the 2001/2002 fishing year. *CCAMLR*-XXII (2003) Report of the Twenty Second Meeting of the Commission, paragraph 8.2.

111 *CCAMLR*-XXI (2002) Report of the Twenty First Meeting of the Commission paragraph 8.74. The proposal was to include William's Ridge, Marion Rise (near Marion Island) and the Del Cano/ Africana Rise taking in statistical areas 51 and 57.

112 Per Sweden and Japan, *CCAMLR*-XXI (2002) ibid, paragraphs 8.79 and 8.82.

113 *CCAMLR*-XXI (2002) Report of the Twenty First Meeting of the Commission paragraph 8.77.

114 Loose document, 'Some thoughts on moving the *CCAMLR* Boundary north ward in the south west Indian Ocean Sector' 2003. (copy on file)

the zone in accordance with *CCAMLR* conservation measures already in force for the lower latitude sub-areas that fall within the Convention Area. The decision to establish a contingency zone could be made without the need to amend Article 1 of the *CCAMLR* which clearly outlines the area of application of the Convention.

The temporary arrangements would stay in place until a suitable RFMO assumed responsibility for managing the marine living resources within area 51.[115] Such action is within the scope of Article 8(5) of the *UN Fish Stocks Agreement* which states:

> Where there is no sub-regional or regional management organisation to establish conservation and management measures for a particular straddling fish stock, relevant coastal States and States fishing on the high seas for such stock in the sub-region or region shall cooperate to establish such an organisation or enter into other appropriate arrangements to ensure conservation and management of such stock and shall participate in the work of the organisation or arrangement.

The proposed contingency zone indicates that the *CCAMLR* Commission is actively fulfilling the role of a competent RFMO within the region. The proposal would also meet the objective of Article XI of *CCAMLR*, which instructs the Commission to seek cooperation with contracting parties exercising jurisdiction in marine areas adjacent to the Convention Area, with a view to harmonising conservation measures adopted in respect of stocks occurring both within the Convention Area and adjacent marine area. In the absence of any RFMO, *CCAMLR*, through the establishment of a contingency zone may facilitate harmonised measures.

Further, adopting a contingency zone would be compatible with Resolutions 10/XII and 17/XX, both of which address stocks occurring outside the Convention Area. Resolution 10/XII, adopted in 1993, is titled *Resolution on Harvesting of Stocks occurring both within and outside the Convention Area*. It recalls the provisions of Article XI of the Convention and reaffirms that:

> Members should ensure their flag vessels conduct harvesting of such stocks in areas adjacent to the Convention Area responsibly and with due respect for the conservation measures it [the Commission] has adopted under the Commission.

Resolution 17/XX was adopted in 2001. It applies specifically to Toothfish located to the north of the Convention Area. The Resolution is aimed at supporting the CDS for both Toothfish species, and in relation to stocks occurring in area 51, urges:

> States participating in the CDS, if necessary to that end, consider reviewing their domestic laws and regulations, with a view to prohibiting, in a manner consistent with international law, landings/transhipments/imports of *Dissostichus spp.* declared in

[115] Ibid and personal Communication with Dr Miller, above.

> a DCD [*Dissostichus* catch documents] as having been caught in FAO Statistical Area 51 if the Flag State fails to demonstrate that it verified the DCD using automated satellite-linked VMS derived data reports.[116]

The membership criteria established by *CCAMLR* also supports its role as competent RFMO under the *UN Fish Stocks Agreement.* As mentioned, membership of the Commission is open, *inter alia*, to:

> Contracting parties which participated in the meeting at which this Convention was adopted…' and to 'acceding parties engaged in research or harvesting activities in relation to the marine living resources to which this Convention applies.[117]

However, States may simply choose to accede to the Convention without seeking Commission membership. In this case they need only demonstrate an interest in research or harvesting activities. Given the vast geographical reach of the Convention and the fact that it applies to all Antarctic marine living resources,[118] this threshold test is not an onerous one to meet. This can be contrasted with the *UN Fish Stocks Agreement* which stipulates that membership of existing RFMOs or arrangements are to be open to States with a 'real interest in the fisheries concerned'.[119] Further, the terms for participation are not to preclude interested States or discriminate against any States or group of States having a real interest in the fisheries concerned.[120] The term 'real interest' is not defined in the *UN Fish Stocks Agreement.*[121] As such, it is not clear whether the term should be construed as imposing a minimum requirement that States actually be engaged in fishing within the relevant regional or sub-regional area. This would impose a higher threshold test than the one required under *CCAMLR.*

---

116 Resolution 17/XX (2001) *Use of VMS and other Measures for the Verification of CDS Catch Data for Areas outside the Convention Area, in particular FAO Statistical Area 51* , paragraph 2. Paragraph 3 refers to the possibility of a metapopulation in that it requests the Scientific Committee to review data concerning the area where *Dissostichus spp.* occur outside the Convention Area and the potential biomass of the *Dissostichus spp.* in such areas, in order to assist the Commission in the conservation and management of *Dissostichus spp.* stocks and in defining the areas and potential biomasses of *Dissostichus spp.* which could be landed/imported/exported under the CDS.

117 *CCAMLR*, Article VII.

118 *CCAMLR*, Article VI, exempts the *International Convention for the Regulation of Whaling* and the *Conservation for the Conservation of Antarctic Seals.*

119 *UN Fish Stocks Agreement*, Article 8(3).

120 Ibid.

121 However, see, E.J. Molenaar, 'The Concept of 'Real Interest' and Other Aspects of Cooperation through Regional Fisheries Management Mechanism' (2000) 15(4) *International Journal of Marine and Coastal* Law, 475, 495. He suggests that one might read into the wording in Article 8(3) an intention to restrict the number of States that may qualify for membership by requiring States participate and cooperate in agreeing upon conservation and management measures. This would exclude States without any real fishing interests but with strong conservation policies. It would also avoid a replication of the membership of the International Whaling Commission where member States advocating a moratorium on whaling out number the pro-whaling members.

Finally at the 2004 Commission Meeting the Chair of the Commission informed delegates that *CCAMLR* had been invited to join the Fishery resource Monitoring System (FIRM) established by the FAO with the aim of drawing together international organisations, regional fisheries bodies, national scientific institutions and the FAO to share information on the status and trends of fishery resources.[122]

5.5 Previous tests to *CCAMLR* Authority

The threat of commercial extinction of the Patagonian Toothfish from overfishing is an overwhelming issue for the *CCAMLR* Commission. It has dominated the agenda for the annual meetings of contracting parties since 1997. In the 24 year history of *CCAMLR*, the rapid escalation of Toothfish catch levels, particularly catches attributable to IUU fishing, heralds the first real challenge to the authority of the Commission.[123] However, the Toothfish is simply the latest in a line of Southern Ocean marine species to be subject to intensive and unregulated harvesting. Following is a brief review of unregulated resource management in the Southern Ocean. It indicates a pattern of exploitation in the Southern Ocean.

In the late 1700's and early to mid 1800's, sealers, apprised of the Southern Ocean's resources by Captain James Cook's voyage reports, targeted fur seal populations on sub-Antarctic islands. From South Georgia, to the South Shetland Island, the South Orkney Islands and Macquarie Island, the sealers systematically slaughtered as many seals as their vessels could hold.[124] Despite the freezing weather and hostile seas, remote sealing and whaling stations were established to support the industry. Indeed, renowned Antarctic explorer Sir Earnest Shackelton owed his life and that of his 22 men to the presence of a whaling station on remote South Georgia Island.[125]

During the early 1820s the South Shetland Islands were surveyed by British sealing ships. Just three years later up to 45 sealing vessels were prowling the shores searching for the declining seal colonies. In 1821, one vessel returned to London with 18,000 seal skins on board.[126] By 1822, the vessels had moved onto Elephant Island and the South Orkney Islands.[127]

---

[122] *CCAMLR*-XXIII (2004) Report of the Twenty Third Meeting of the Commission, paragraph 14.27.

[123] 15,931 tonnes of Toothfish were harvested from within the Convention Area during the 2003/2003 fishing year. A further 18,919 tonnes were reported to be harvested from the high seas area outside and adjacent to the Convention Area. *CCAMLR*-XXII (2003) Report of the Twenty Second Meeting of the Commission, paragraphs 4.40-4.41. See also, Table 1.2 in Chapter 1, which illustrates the percentage of reported catches attributable to IUU fishing over recent years.

[124] G. E Fogg, A *History of Antarctic Science* (1992) 39. See also, Map 4.1.

[125] R. Huntford, *Shackelton* (2000) 597. Shackelton and two of his men arrived at Stromness Bay on South Georgia Island on 20 May 1926.

[126] A. Gurney, *Below the Convergence* (1997) 191-193

[127] Ibid.

In the late 19th century, whalers, faced with depleted whale populations in northern hemisphere waters, sent exploratory expeditions to the Southern Ocean.[128] During the mid to late 1900s, whales were systematically fished with such intensity that the blue whale, humpback whale and the southern right whale were listed as endangered species under the *Convention on International Trade in Endangered Species* (CITES).[129] During the 1930-31 fishing season alone, over 31, 000 blue whales were slaughtered, an assault from which the species has never recovered.[130]

After the whales had been commercially depleted, attention turned to finfish. Antarctic finfish were first harvested during the 1969 season by the Soviet fishing fleet.[131] For some time, a number of fin fish species were targeted, including the marbled notothenia (*Notothenia rossi)*, mackerel icefish (*Champsocephalus gunnari*), the grey notothenia (*Notothenia squamifrons*) Gunther's notothenia (*Patagonotothen guntheri*), the sub-Antarctic laternfish (*Electrona carlsbergi*) and the Patagonian Toothfish (*Dissostichus eleginoides*).[132] From South Georgia Island the high seas fishing fleet moved west the following season towards the Kerguelen Isles, a French possession since its discovery by Captain Yves de Kerguelen in 1772. Other nations, including Poland, the then German Democratic Republic and Bulgaria, joined the fishery and finfish continued to yield good catches throughout the 1970s.[133]

Also at this time, the Soviet high seas fishery adjacent to the Kerguelen Isles started to raise concerns for the French authorities. Subsequently, an EEZ was declared for the French Southern and Antarctic Territories in 1978.[134] The AFZ was declared the following year and included waters adjacent to the Heard and MacDonald Islands.[135] Commercial long lining for the Patagonian Toothfish commenced during the 1990-1991 season when Chile, initially with just one vessel, fished the waters adjacent to South Georgia.[136]

128 Fogg, *A History of Antarctic* Science above, 106. The right whale had by that time had been harvested to the point of extinction.

129 Opened for signature 3 March 1973, 12 ILM 1085 (entered into force 1 July 1975). ('CITES').

130 Gurney, *Below the Convergence* above, 61

131 Kock, 'Fishing and Conservation in the Southern Ocean, above, 4.

132 Ibid.

133 Ibid.

134 Ibid.

135 See Map 8.1.

136 Kock, 'Fishing and Conservation in the southern Ocean' above, 4-5.

5.6 The development of the Patagonian Toothfish Fishery

Illegal fishing was described in 2001 as 'one of the greatest challenges the Commission has had to face'.[137] Yet only a decade earlier, the statement was made that:

> [C]ontrary to the problem faced by other fisheries commissions, no threat to the conservation of Antarctic marine living resources is currently posed by fishing operations of non contracting parties or vessels re-flagged to flag of convenience states.[138]

5.6.1 The Patagonian Toothfish – Statistics and Habitat

There are actually two species of Toothfish in the Southern Ocean. Both are similar in appearance and habit, however the Antarctic Toothfish (*Dissostichus mawsoni)* is found closer to the Antarctic continent where sea ice forms and has not yet been subject to harvesting in the same quantities as the Patagonian Toothfish.[139]

The Patagonian Toothfish is a demersal, or sea-floor dwelling species. However at some stages of life it is pelagic (meaning living at or near the ocean surface).[140] It is a large fish reaching up to 2.2 metres in length and up to 100 kilograms in weight.[141] They can be found in waters from 300 to 2,500 metres deep though the depths do not protect them. Longline systems have been developed to target the bigger fish dwelling at 2,500 metres.[142] The species is vulnerable to the impacts of overfishing for two reasons. First, it matures very slowly and takes 10-12 years to reach spawning age. Scientific knowledge on its maximum age is uncertain, although the maximum recorded age is 45 years.[143] Secondly, due to the recent discovery and remote location of the fishery, there have been difficulties in obtaining reliable scientific data about the dispersal of stock and population groupings, which in turn hampers stock management.[144] Unregulated and unreported fishing compounds this fact.

---

137 'Statement for the Commemoration of 20 years of *CCAMLR*', above.

138 Ibid.

139 Australian Antarctic Data Centre, 'Toothfish:10 Facts' (undated). See, also, D.J. Agnew, 'The Illegal and Unregulated Fishery for Toothfish in the Southern Ocean and the *CCAMLR* Catch Documentation Scheme' (2000) 24 *Marine Policy* 316, 361.

140 AFMA Fact Sheet, 'Commonwealth Fisheries, Heard and McDonald Islands – Principal Commercial Fish Species' undated.

141 Ibid.

142 ISOFISH, *The Involvement of Mauritius in the trade in Patagonian Toothfish from illegal and unregulated long line fishing in the Southern Ocean and what might be done about it.* Occasional Report No.1 (August 1998 3rd ed) Part 4.2.

143 Australian Antarctic Data Centre 'Toothfish : 10 Facts' above.

144 AFMA Fact Sheet 'Principal Commercial Fish Species' above. See also discussion about the meta-population in the Indian Ocean at CCAMLR-XXIII (2003) Report of the Twenty third Meeting of the Commission, paragraphs 11.2-11.3.

The Toothfish is a valued commodity in both the Japanese and American markets. It is traded under a variety of names. On the Asian market it is known as the *Mero.* It is sold as *Chilean Seas Bass* or simply *Sea Bass* in the USA. In southern Chile, fishermen refer to it as *Merlusa nigra* (black hake).[145] The range of names can act against efforts to track the illegal trade in Toothfish.[146] Prices can vary from US$5,000-7,000 per tonne,[147] making it an attractive catch for coastal and high seas fishing States alike.

The larger, more valuable stocks are found within the sub-Antarctic exclusive economic zones of South Africa, France, Australia and New Zealand.[148] Significantly, the Toothfish has been identified as a straddling fish stock. That is one which is located both within coastal State waters and areas of high seas. In 1992, the Scientific Committee first raised the possibility of the Toothfish being a straddling fish stock.[149] In 1995, the Working Group on Fish Stocks Assessment confirmed that the Toothfish population found in waters adjacent to South Georgia was a straddling Stock.[150]

As discussed above, the Toothfish is also a transboundary fish stock. That is it occurs both within and outside the *CCAMLR* Convention Area. The management of transboundary stocks falls under the same framework as straddling stocks. In 1993 the Commission referred to the transboundary distribution pattern of the Toothfish by acknowledging that 'stocks occur both within and outside the Convention Area'[151] In 2003 the debate on the extent of the transboundary population is still open.[152]

The Toothfish is known to exist in good quantities on the Kerguelen Plateau which is a continental shelf upon which the Kerguelen Isles and Heard and McDonald Island are situated. Commercially valuable populations straddle areas of high seas adjacent to both the French exclusive economic zone around the Kerguelen Isles and the AFZ declared offshore to the Heard and McDonald Islands.[153] Portions of these adjacent high seas areas fall within the *CCAMLR* Convention Area. As discussed, this fact does not alter their status as high seas for non *CCAMLR* members.

---

145 ASOC Press Release, 'Patagonian Toothfish- Going to hell in a fishing basket' 29 October 1999.

146 ISOFISH *The Involvement of Mauritius in the trade in Patagonian Toothfish,* Occasional Report No.1 above.

147 AAP Press Release (001 PAC) 'Valuable Fish Species being plundered in the Southern Ocean' 18 May 1997.

148 Ibid.

149 *CCAMLR*-XIII (1992) Report of the Thirteenth Report of he Commission, paragraph 143.

150 *CCAMLR*-XIV (1995) Report of Scientific Committee, Annex 5, paragraph 10.11.

151 *CCAMLR*-XIV (1993) Report of the Fourteenth Meeting of the Commission, 57-58.

152 Above n 4 8.

153 See Maps 6.2 and 6.3 in Chapter 6.

5.6.2 Discovery of the Toothfish

The Patagonian Toothfish fishery has recent beginnings.[154] The development of long line fishing around South Georgia from the late 1980s marked the start of large-scale commercial fishing.[155] A number of States, including *CCAMLR* members, commenced harvesting in the south-west Atlantic Ocean around South Georgia.[156] In response, the *CCAMLR* Commission set a total allowable catch of 2,500 tonnes for the South Georgia statistical area during the 1990-1991 fishing season.[157] The total allowable catch for 1991-1992 was increased to3,500 tonnes.[158]

After exhausting the fishery around South Georgia, vessels moved eastward via Bouvet Island; Prince Edward and Marion Island; Crozet and the Kerguelen Islands; before reaching the Heard and McDonald Islands.[159] The fishery is developing in one of the most remote areas of ocean in the world. For this reason *CCAMLR* is particularly vulnerable to the effects of IUU fishing.[160] Remote and isolated seas are a haven for illegal fishers. Consistent high sea states and extreme weather conditions make effective enforcement a logistical challenge.

In 1995, Australia and South Africa advised the Commission of their respective intentions to commence a new fishery within their coastal State waters that fell within the Convention Area.[161] A total allowable catch for sub-area 58.5.2, the Heard and McDonald Islands' area, was established at the precautionary limit of 297 tonnes.[162] At the 2003 *CCAMLR* meeting, the Commission endorsed advice from the Scientific Committee by adopting a catch limit of 2,873 tonnes for area 58.5.2.[163]

5.6.3 Concerns about IUU fishing

Concerns were first raised in 1993 about illegal fishing activities in the Southern Ocean. The *CCAMLR* Scientific Committee reported that Toothfish stocks around South Georgia Island may have been depleted to approximately 30% of the original stock levels.[164]

---

[154] However, Chile is reported to have investigated exploratory fishing as early as 1955. Agnew, 'The Illegal and Unregulated Fishery for Toothfish' above, 361, 362.

[155] Ibid and see Kock , 'Fishing and Conservation in the southern Ocean' above.

[156] Chile reported a catch of 3,012 tonnes and Russia 598.1 tonnes for the South Georgia area.

[157] *CCAMLR* –IX (1990) Report of the Ninth Meeting of the Commission, paragraph 13.37. CM24/IX. The *CCAMLR* fishing year then ran from 1November 1990 to 1 November 1991. It now runs from 1 December to 30 November.

[158] *CCAMLR* –X (1991) Conservation measure 35/X.

[159] ISOFISH, 'Patagonian Toothfish Profile' undated.

[160] E.J. Molenaar, ' *CCAMLR* and Southern Ocean Fisheries' (2001) 16(3) *International Journal of Marine and Coastal Law* 465, 498.

[161] *CCAMLR* –XIV (1995) Report of the Scientific Committee paragraphs 8.1, 8.7.

[162] *CCAMLR* XIV (1995) C78/XIII amended to conservation measure 78/XIV.

[163] *CCAMLR*-XXII (2003) Report of the Twenty Second meeting of the Commission, paragraph 10.52.

[164] *CCAMLR* –XII (1993) Report of Scientific Committee paragraphs 3.34-3.38. Note the word 'may'

Signs of active management appear in the agenda for the 1995 *CCAMLR* annual meeting. The Working Group on Fish Stock Assessment reported the 'unreported catch was either of the same order or higher than the reported catch.'[165] It was also observed that over the previous four years, the reported catch represented just 40% of the estimated total catch harvested from within sub-area 48.3 (South Georgia Island) and adjacent areas.[166]

It is difficult to judge estimates of the quantity of Toothfish harvested by the IUU fishing industry. However IUU fishing as a percentage of the total catch of Toothfish within the Convention Area is significant. Table 1.1 in Chapter 1 illustrates the extent of the problem raised by IUU fishing in terms of reporting the rising IUU catch. Whilst the peak of IUU fishing appears to have momentarily passed, it remains a significant management problem for the *CCAMLR* Commission. Contracting party vessels remain involved in IUU fishing, with the report that two contracting party vessels were added to the 'IUU Vessel List of Contracting Parties in 2004.[167] Furthermore, discussion at the 2004 *CCAMLR* Commission meeting indicate that contracting party vessels have been involved in providing tampered Vessel Monitoring System (VMS) data to the *CCAMLR* Secretariat.[168]

Instances of noncompliance with conservation measures were noted by the Commission during the 1995 *CCAMLR* Commission meeting. For example, a South Korean vessel (contracting party State) was reported for conducting fishing operations out of season within the South Georgia statistical area.[169] Further, Belize, a non-contracting party State, responded to representations regarding the sighting of a Belize flagged fishing vessel within the South Georgia Island sub-area by assuring the Commission that the vessel would not return. The vessel was subsequently sighted within the same area during the same fishing season.[170]

In 1996 member States expressed further concerns about fishing without regard to relevant conservation measures by vessels flagged to Argentina, Belize, Panama, Vanuatu and Portugal.[171] South Africa reported that four vessels were detected fishing within the South African EEZ around the Prince Edward Islands or adjacent *CCAMLR* waters, during a surveillance flight in September 1996.[172] Two of the

---

was used as there was in 1993 and remains today, uncertainty about exact population levels. See also Agnew, 'The Illegal and Unregulated fishery for Toothfish' above, 361, 362.

165 *CCAMLR* XIV (1995) Report of the Twenty Fourth Meeting of the Commission, 302.

166 Ibid, 276.

167 *CCAMLR*-XXIII (2004) Report of the Twenty Third Meeting of the Commission, paragraph 8.3.

168 *CCAMLR*-XXIII (2004) Illegal Toothfish Trade: Introducing Illegal Catches into the Markets' 7. Copy accessed with permission of the author, ASOC. Copy held on file.

169 *CCAMLR* –XIV (1995) Report of Standing Committee on Observation and Inspection (SCOI) 121.

170 Ibid, 125.

171 *CCAMLR* –XV (1996) Report of Standing Committee of Observation and Inspection, paragraphs 1.26, 1.53. Argentina and Portugal (through EC membership) are members of the Commission. Vanuatu has acceded to the Convention. Belize and Panama are known flag of convenience States.

172 Ibid, paragraphs 1.39-1.40.

vessels sighted had, reportedly, been originally flagged to *CCAMLR* contracting party States.[173] This information is supported by reports of substantial illegal catches being landed at Port Louis (Mauritius), Walvis Bay (Namibia) and in port in Mozambique, between September 1996 and March 1997.[174] The estimates placed the unreported catch between 74,000 and 82,200 tonnes.[175] Concerns that fish stocks were being over harvested are also supported by reports that fish landed from waters around Prince Edward Island in 1997 had dropped, below the estimated spawning length, to 60cm in length.[176]

Early press coverage of the developing IUU fishing industry commenced in 1997. On 30 April 1997, it was reported that France and Britain had sent warships to the Southern Ocean to warn off the illegal fishing vessels.[177] New Zealand was also reported to be cooperating with South Africa, France and Australia to halt the illegal fishing.[178] Concerns emerged on the Australian domestic front with reports of up to 70 illegal vessels operating within Australia's remote fishing zone adjacent to the Heard and McDonald Islands.[179] At that time the Australian Government described the remote fishing zone as being 'at the maximum extreme of its fisheries protection services.' [180]

Early arrests were also reported. In March and April 1997, the French Navy arrested two vessels, for allegedly fishing illegally for Toothfish, off Crozet Island.[181] A pattern of moving eastward across the Southern Ocean, like a slow moving plague had commenced. During discussions at the 1997 *CCAMLR* meeting many member States expressed concern at the growing incidence of illegal fishing within the Convention Area. The European Community (EC) stated that it: 'considered that *CCAMLR* faced a major challenge resulting from the blatantly illegal and/or non-notified fishing activities.'[182] Further, by the *CCAMLR* Commission, was considered by the EC, to be 'at risk, not only of being undermined, but irreparably damaged by these activities.'[183] At the 1987 meeting, the Commission concluded, that fisheries management is aggravated by the fact:

> [T]hat illegal, unregulated fisheries and unreported catches today exceed reported fishing by a factor several times over. No less

---

[173] Ibid, paragraph 1.41.

[174] *CCAMLR* XVI (1997) Report of Working Group on Fish Stock Assessment. Annex 5, 239-425.

[175] Ibid.

[176] Ibid.

[177] AP/AAP Press Release (OO2PAC) 'Opening round of battle for Toothfish in Antarctica' 30 April 1997.

[178] Ibid.

[179] AFP Press Release, 'Valuable Fish Species being plundered in the Southern Ocean' 18 May 1997.

[180] Ibid.

[181] Ibid.

[182] *CCAMLR* XVI (1997) Report of the Sixteenth Meeting of the Commission, paragraph 5.2.

[183] Ibid.

> aggravating is the fact that more than half of the vessels presumed to engage in illegal, unregulated and unreported fishing fly the flags of *CCAMLR* Member States. That underlines the urgent need for *CCAMLR* and *CCAMLR* Member States to bring their own house in order.[184]

Concerns over illegal and unregulated fishing were expressed once more at the 1998 *CCAMLR* meeting. A few months prior to the meeting, the Australian Government made the following statement:

> If illegal and unregulated fishing continues at the current level the population of the Patagonian Toothfish will be so severely decimated that within the next two to three years the species will be commercially extinct. Some areas are already showing signs of this.[185]

The 1998 Report of the Standing Committee on Observation and Inspection noted that there were 45 reported sightings of non-member fishing vessels within the Convention Area.[186] Reports of non-compliance by member State flagged vessels were also noted. Two Chilean flagged vessels were detained within the French exclusive economic zone offshore from the Kerguelen Islands. A Belize flagged vessel with ownership ties to a Uruguayan company was arrested within the Australian fishing zone.[187] The problem of reflagging to 'flag of convenience' States was emerging as a more common issue.

Noting, in 1998, that IUU fishing was comparable to a 'cancer eating at the fibre of the *Antarctic Treaty* System'[188] contracting party States agreed that the 'most stringent measures possible'[189] should be taken to combat the problem. In 2002 the Standing Committee on Observation and Inspection observed that IUU fishing 'was now a highly organised form of transnational crime.'[190]

Recognising the problem is one thing. Responding effectively quite another. The following Chapter reviews the measures and resolutions adopted by the *CCAMLR* Commission to address the threat of IUU fishing.

---

184 Ibid, paragraph 5.8.

185 Minister for Resources and Energy and Minister for Foreign Affairs and Trade, Press Release, 'Australia welcomes Norwegian Government action on Illegal Fishing' FA98, 22 July 1998. More recent estimates suggest anywhere from 2-4 years . J.Vidal, 'Southern Ocean hunt for ship with cargo of endangered Toothfish' The Guardian (London) 19 August 2003, 11, per the World Conservation Union- Traffic and the US National Environmental Trust.

186 *CCAMLR* XVII (1998) Report of Standing Committee on Observation and Inspection, paragraph 2.2.

187 Ibid, paragraphs 2.9-2.11.

188 *CCAMLR* XVII (1998), paragraph 5.9.

189 Ibid. paragraph 5.12.

190 *CCAMLR*-XXI (2002) Annex 5, paragraph 5.3.

Annex A to Chapter 5

Statement by the Chairman of the Conference on the Conservation of Antarctic Marine Living Resources

The Conference on the Conservation of Antarctic Marine Living Resources decided to include in the publication of the Final Act of the Conference the text of the following statement made by the Chairman on 19 May 1980 regarding the application of the Convention on the Conservation of Antarctic Marine Living Resources to the waters adjacent to Kerguelen and Crozet over which France has jurisdiction and to waters adjacent to other islands within the area to which this Convention applies over which the existence of State sovereignty is recognised by all Contracting Parties.

1. Measures for the conservation of Antarctic marine living resources of the waters adjacent to Kerguelen and Crozet, over which France has jurisdiction, adopted by France prior to the entry into force of the Convention, would remain in force after the entry into force of the Convention until modified by France acting within the framework of the Commission or otherwise.
2. After the Convention has come into force, each time the Commission should undertake examination of the conservation needs of the marine living resources of the general area in which the waters adjacent to Kerguelen and Crozet are to be found, it would be open to France either to agree that the waters in question should be included in the area of application of any specific conservation measure under consideration or to indicate that they should be excluded. In the latter event, the Commission would not proceed to the adoption of the specific conservation measure in a form applicable to the waters in question unless France removed its objection to it. France could also adopt such national measures as it might deem appropriate for the waters in question.
3. Accordingly, when specific conservation measures are considered within the framework of the Commission and with the participation of France, then:

   (a) France would be bound by any conservation measures adopted by consensus with its participation for the duration of those measures. This would not prevent France from promulgating national measures that were more strict than the Commission's measures or which dealt with other matters;

   (b) in the absence of consensus, France could promulgate any national measures which it might deem appropriate.
4. Conservation measures, whether national measures or measures adopted by the Commission, in respect of the waters adjacent to Kerguelen and Crozet, would be enforced by France. The system of observation and inspection foreseen by the Convention would not be implemented in the waters adjacent

to Kerguelen and Crozet except as agreed by France and in the manner so agreed.

5. The understandings, set forth in paragraphs 1 to 4 above, regarding the application of the Convention to waters adjacent to the islands of Kerguelen and Crozet, also apply to waters adjacent to the islands within the area to which this Convention applies over which the existence of State sovereignty is recognised by all Contracting Parties.

enhancing the effectiveness of the Scheme in regulating IUU fishing. However, the *CCAMLR* Commission has also witnessed the development of new port States such as Indonesia and Mozambique, both of which have been approached to implement the CDS.[14]

It has been noted as recently as 2003 that 'given the obvious legal limitations of measures addressed to the flag State, it is somewhat surprising that port State controls have not been more readily adopted as a means to deter IUU fishing.[15] In the author's view RFMOs such as *CCAMLR* and *ICCAT* are seeking to extend the culture of port State control from IMO sanctioned safety inspections to high seas and regional regulation. The success of the *CCAMLR* CDS suggests that port State controls will be more readily supported and adopted in the future, particularly if flag States continue to be recalcitrant in discharging their duties in relation to fishing vessel regulation.

The role of some port Sates in facilitating the IUU trade in Toothfish was acknowledged by the *CCAMLR* Commission in 1997 when it resolved that:

> [T]he Commission consider inviting as observers to its next meeting, some States known to be providing, perhaps unwittingly, port access and other services to vessels implicated in illegal and unregulated fishing in the Convention area.[16]

The *CCAMLR* Secretariat acted decisively by inviting both Mauritius and Namibia to attend, as observers, the 1998 *CCAMLR* Commission meeting.[17] At that time large quantities of Toothfish were moving through Walvis Bay in Namibia and Port Louis in Mauritius. It was estimated that tens of thousands of tonnes of Toothfish were being processed annually through Port Louis.[18] Given that Mauritius exported Toothfish (and products) directly to the principal the markets of Chile, Hong Kong, Japan, Singapore, Tawian and the USA, securing the cooperation of the Mauritian government in implementing port State controls became a priority for the *CCAMLR* Commission.[19]

In 2000, the Acting Prime Minister of Mauritius acknowledged the role his State had played in the facilitation of IUU fishing in the Southern Atlantic and pledged that the new government would 'cooperate fully to fight illegal fishing of Toothfish.

---

14 *CCAMLR*-XXII (2003), Report of the Twenty Second Meeting of the Commission, paragraphs 8.8-7.9.

15 C. Hedley and L. Kengkran, 'FAO Expert Consultation on Port State Measures to combat IUU Fishing' [2003] *International Fisheries Law and Policy Review* (Special Issues) 32.

16 *CCAMLR* –XVI (1997) Report of the Sixteenth Meeting of the Commission, paragraphs 5.36-5.37.

17 Ibid.

18 ISOFISH, *The Involvement of Mauritius in the trade in Patagonian Toothfish from Illegal, Unregulated Longline Fishing in the Southern Ocean and what might be done about it*, Occasional Report No.1 (1988, 3rd ed) Section 1.2.

19 Greenpeace, 'Southern Ocean Pirate Fishing Expedition' (2000) <www.greenpeace.org/pirate fishing> (15 May 2001).

through the medium of RFMOs, has evolved from a treaty obligation to become a matter of customary international law is well grounded. It has recently been stated 'that it is arguable that States may now accept, as a matter of customary international law, the obligation to cooperate in respect of high seas fisheries through the medium of RFOs or other cooperative arrangements.'[129] Prior to this in 1994, Burke observed that the obligation 'is probably considered also to be a customary law of the sea principle.'[130] In support of this statement, he referred to the ICJ decision in the *Fisheries Jurisdiction Case*,[131] in which the ICJ found that the parties 'were required to examine together in light of the scientific and other available information, the measures required for the conservation and development and equitable exploitation' of resources in the high seas beyond 12 miles.'

The duty to cooperate under the *LOCS* was judicially considered and affirmed by ITLOS in the *Southern Bluefin Tuna cases*.[132] This case involved an ongoing dispute between Australia and New Zealand with Japan[133] over the limit set for the total allocated catch and the legality of Japan's experimental fishing program to test the recovery of the tuna stocks at various places.[134] Of relevance to this study, Australia and New Zealand argued that Articles 64, and 117-119 of the *LOSC* imposed a duty on Japan to cooperate in the conservation and management of high seas resources.[135] In granting Provisional Measures, ITLOS considered that there was a duty to cooperate directly or through appropriate international organisations with a view to ensuring conservation and promoting the objective of optimum utilisation of highly migratory species.[136] Further, ITLOS considered that the conduct of the member parties to the *Commission for Conservation of Southern Bluefin Tuna* as

129 R.Rayfuse, ' To our Children's Children's Children: From Promoting to Achieving Compliance in High Seas Fisheries' (paper presented to International Conference on the Governance of High Seas Fisheries and the United Nations Fish Agreement, St John's, Newfoundland and Labrador, 1-5 May 2005, 10). <www.dfo-mpo.gc.ca/fgc-cgp/documents/rayfuse_e.htm> (14 May 2005).

130 W.T.Burke, *The New International Law of Fisheries: UNLOS 1982 and Beyond* (1994) 121-122.

131 *UK v Iceland* (1974) ICJ Rep. 31.

132 *New Zealand v. Japan* and *Australian v. Japan* (*Southern Bluefin Tuna Cases*) (Order for Provisional Measures) ITLOS Case Nos. 3 and 4, 27 August 1999.

133 This dispute between Australia and New Zealand, and Japan arose out the unilateral declaration by Japan of a three year experimental fishing program for Southern Bluefin tuna setting a total allowable catch of 1464 tonnes per annum. For the background to the dispute including the formation of the 1993 *Convention for the Conservation of Southern Bluefin Tuna* see C. Romano, 'The Southern Bluefin Tuna Dispute: Hints of a World to Come...Like it or Not' (2001) 32 *Ocean Development and International Law* 313, 315-317 and B, Kwaitkowska 'The Australian and New Zealand v. Japan Southern Bluefin Tuna (Jurisdiction and Admissibility) Award of the First Law of the Sea Convention Annex VII Arbitral Tribunal' 16(2) (2001) *International Journal of Marine and Coastal Law* 239. The Southern Bluefin Tuna was managed under the *Convention for the Conservation of Southern Bluefin Tuna* 1993, opened for signature 1993, 26 Law of the Sea Bulletin 57 (entered into force on 20 May 1994).

134 Romano, 'The southern Bluefin Tuna Dispute' Ibid, 316.

135 Australian and New Zealand Statement of Claim, 15 July 1999.

136 *Southern Bluefin Tuna Cases*, above, paragraph 48.

between themselves, and in interaction with non-member States, was to be judged with reference to their obligations under the *LOSC*.[137]

Whilst the order of ITLOS is encouraging, it was only for Provisional Measures and these Measures were revoked by the Arbitral Tribunal, established under Article 287(3) of the *LOSC*. The Arbitral Tribunal found, by majority, that it had no jurisdiction to hear claims by Australia and New Zealand and Japan.[138] This was a missed opportunity to further develop jurisprudence in the management of high seas fisheries. However the approach taken by ITLOS in awarding Provisional Measures cannot be disregarded and arguably has had an impact on high seas management.

In a recent dispute between Chile and the European Union relating to the migratory swordfish stocks, Chile argued that the EU had breached the duty to cooperate with the coastal State as required under Articles 64 and 117-119 of the *LOSC*.[139] Whilst this case has been suspended on the request of the parties,[140] it is argued that had the Tribunal been given the opportunity to consider the merits of the dispute, its previous order in the *Southern Bluefin Tuna Case*, together with previous ICJ decisions, would feature strongly in the Tribunals findings. In this regard it is considered that when the Tribunal is provided with the opportunity to consider the nature of the obligation to cooperate, it will find that the duty does exist.[141] What is unknown is the extent to which guidance will be given by ITLOS on how that duty is to be discharged.

4.8.5 Market State or Trade Controls

Market State or trade controls should be regarded in conjunction with port State controls. Whilst port State controls can create logistical difficulties for IUU operators as they are forced to seek alternative ports further away from their chosen fishing ground, market controls can operate as a second line of defence preventing the trade in IUU fish even if is successfully landed and unloaded. Port States can deny the right to land fish. However, assuming that IUU operators will persevere until they visit a port that will allow landing, market State or trade controls to prevent the export or import of IUU fish, become more important.

Whilst the *Code of Conduct* does refer to general guidelines for responsible international trade,[142] the concept was not expanded upon until discussions

---

137 *Southern Bluefin Tuna Cases*, above, paragraph 50.

138 *Decision of Arbitral Tribunal, Southern Bluefin Tuna* Case 2000, 39 ILM 1359, 1393.

139 *Case concerning the Conservation and Sustainable Exploitation of Swordfish Stocks and the South-Eastern Pacific Community* (*Chile v the European Union*) ITLOS Case No.7.

140 See ITLOS Press Release No. 87, 7 January 2004.

141 On this point, it is noted that the ICJ in the *North Sea Continental Shelf Case* (1969) ICJ Rep. 4, observed that negotiations should be meaningful. Further, in order for a practice to become a matter of customary international law there must also be evidence of State belief in the practice or *opinio juris*.

commenced for the formation of the *IPOA-IUU*. The relevant provisions of the *IPOA-IUU* have been discussed above.

The FAO has estimated that 'more than 800 fish species are traded internationally in many forms, shapes, brands and preparations.'[143] At least 196 States were involved in exporting 22 million tonnes of fish in 1996. Approximately 180 States imported this fish.[144] The requirement for market State or trade controls to be WTO consistent is examined in detail in Part 6 following the review of *CCAMLR* measures.

International trade is conducted within the framework of the WTO.[145] Whilst the purpose of the WTO is the liberalisation of trade or the furtherance of free trade amongst States, it has been acknowledged that there are instances where it is appropriate that States impose trade restrictions to protect the environment. The Committee on Trade and Environment was established in 1994 to identify the relationship between trade and the environment.[146] The Committee's aim is to make international trade and environmental policies mutually supportive although achieving this in practice is not a straight forward matter.[147]

The fluid nature of IUU fishing has seen an increase in use of re-supply and transshipment vessels to facilitate IUU fishing operations and avoid around port and market State controls. Transshipping catches allows the IUU fishing vessel to remain in situ. The transshipment vessels can then seek ports outside the relevant RFMO area of application which do not apply agreed port State control measures or mix the IUU catch with legal catches and seek to land the entire catch with false catch documentation. This fact alone makes it essential to apply port and market State controls uniformly and consistently as has been done in relation to the safety of shipping and pollution controls under the IMO port State control system of regional MoUs.

4.8.6 National Measures

The *IPOA-IUU* is the only international instrument to focus on the measures that may be implemented by national governments, including coastal States, to deter IUU fishing. As such it provides a reference point for individual States seeking to assist

142 The requite *opinio juris* can be gleaned from the 50 or more RFOs existing to manage the world's marine fisheries. These RFO are evidence of a commitment to and belief in the need to cooperate, even if the achievement of agreed measures is not always realised.

143 *Code of Conduct*, Article 11.2 and 11.3.

144 FAO, The State of World Fisheries and Aquaculture (2004) 82.

145 S.Bache, M.Haward and S.Dovers, 'The Impact of Economic, Environmental and Trade Instruments upon Australian Fisheries policy and Management' (2000) Report prepared for the Fisheries and Aquaculture Branch, Australian DepArticlement of Agriculture, Fisheries and Forestry. (copy on file).

146 Established on 1 January 1995. Before the WTO international trade was conducted under the General Agreement on Tariffs and Trade (GATT) 1947.

147 Ministerial Declaration on Trade and the environment adopted on 1 April 1994 calling for the establishment of the Committee on Trade and Environment.

in the international and regional response to IUU fishing. As will be demonstrated in the review of Australian action against IUU fishers in Part 5, coastal State or national action complements the global response to IUU fishing. The ways in which national governments can respond are undertaken voluntarily however this fact does not diminish their value as legal measures to deter IUU fishing under international law. They are in accordance with existing principles of international law and can be very effective in addressing the economic 'drivers' of IUU fishing. Moreover, if implemented on a co-operative bilateral basis the impact is further enhanced.

4.9 Conclusion

This Chapter has examined the response of the international community in establishing binding treaties, frameworks, international plans and normative standards to deter IUU fishing. Following the review of individual conventions, agreements and plans, Section 1.10 of this Chapter draws the relevant international law together in the context of the measures which may be implemented and analyses them on the basis of either treaty or customary internal law obligations. There has been no comprehensive review of the state of international law in relation to measures to deter IUU fishing. Whilst IUU vessel lists and port and market State controls remain matters to be incurred through treaty obligations, it is argued that the duty to cooperate in the management of high seas fisheries has emerged as a matter of customary international law. This fact is encouraging and important for it is through the implementation of port and market State controls, the compilation of IUU vessel lists and the exchange of information that this duty to cooperate may be discharged.

The following Chapters introduce the management of Southern Ocean fisheries first at the regional level under *CCAMLR* and then under national legislation by Australia. It is in these two Parts that the importance of analysis undertaken in this and the preceding Chapters will become evident. Measures taken by both the *CCAMLR* Commission and Australia take account of both the factors contributing to the development of IUU fishing and the current international law addressing the activity.

Annex A to Chapter 4

Agreement to Promote Compliance with International Conservation and Management Measures by Fishing Vessles on the High Seas

Participant	Acceptance
Argentina*	24 Jun 1996
Australia*	19 Aug 2004
Barbados	26 Oct 2000
Benin	4 Jan 1999
Canada*	20 May 1994
Chile*	23 Jan 2004
Cyprus	19 Jul 2000
Egypt	14 Aug 2001
European Community*	6 Aug 1996
Georgia	7 Sep 1994
Ghana	12 May 2003
Japan*	20 Jun 2000
Madagascar	26 Oct 1994
Mauritius	27 Mar 2003
Mexico	11 Mar 1999
Morocco	30 Jan 2001
Myanmar	8 Sep 1994
Namibia*	7 Aug 1998
Norway*	28 Dec 1994
Peru*	23 Feb 2001
Republic of Korea*	24 Apr 2003
St. Kitts & Nevis	24 Jun 1994
St. Lucia	23 Oct 2002
Seychelles	7 Apr 2000
Sweden*	25 Oct 1994
Syrian Arab Republic	13 Nov 2002
Tanzania	17 Feb 1999
United States of America*	19 Dec 1995
Uruguay*	11 Nov 1999

(29 member States)

Agreement for the Implementation of the Provisions of the Convention Relating to the Conservation and Managment of Staddling Fish Stocks and Highly Migratory Fish Stocks [148]

Participant	Acceptance
Kenya	13 July 2004
Sweden*	19 December 2003
Spain*	19 December 2003
Portugal	19 December 2003
Netherlands*	19 December 2003
Luxembourg	19 December 2003
Italy*	19 December 2003
Ireland	19 December 2003
Greece*	19 December 2003
Germany*	19 December 2003
France *	19 December 2003
Finland*	19 December 2003
Denmark	19 December 2003
Belgium*	19 December 2003
Austria	19 December 2003
European Community*	19 December 2003
India*	19 August 2003
South Africa*	14 August 2003
Marshall Islands	19 March 2003
Ukraine*	27 February 2003
Cyprus	25 September 2002
United Kingdom* on behalf of Pitcairn, Henderson, Ducie and Oeno Islands, Falkland Islands, South Georgia and South Sandwich Islands, Bermuda, Turks and Caicos Islands, British Indian Ocean Territory, British Virgin Islands and Anguilla	10 December 2001
United Kingdom of Great Britain and Northern Ireland (on behalf of the United Kingdom of Great Britain and Northern Ireland)	19 December 2003
Malta	11 November 2001
Costa Rica	18 June 2001
New Zealand*	18 April 2001
Barbados	22 September 2000
Brazil*	8 March 2000

Australia*	23 December 1999
Uruguay*	10 September 1999
Canada*	3 August 1999
Monaco	9 June 1999
Papua New Guinea	4 June 1999
Cook Islands	1 April 1999
Maldives	30 December 1998
Islamic Republic of Iran	17 April 1998
Namibia	8 April 1998
Seychelles.	20 March 1998
Russian Federation	4 August 1997
Federated States of Micronesia	23 May 1997
Mauritius	25 March 1997
Iceland	14 February 1997
Solomon Islands	13 February 1997
Senegal	30 January 1997
Bahamas	16 January 1997
Nauru	10 January 1997
Norway	30 December 1996
Fiji	12 December 1996
Samoa	25 October 1996
Sri Lanka	24 October 1996
United States of America	21 August 1996
Saint Lucia	9 August 1996
Tonga	31 July 1996

(52 member States)
[* in both tables denotes *CCAMLR* member]

148 Refer WTO, 'Committee on Trade and Environment', 6. <www.wto.org.>

149 Source http://www.un.org/Depts/los/reference_files/chronological_lists_of_ratifications.htm

PART 4

Regional Measures: *CCAMLR*

Moreover he welcomed the French offer of assistance extended to the Mauritian team conducting investigations of fishing vessels calling at the Port Louis Harbour.'[20] The Mauritian government further stated that it would examine a number of options to deter IUU fishing of Toothfish including adopting the *CCAMLR* CDS (which it now does) and closing Port Louis to fishing vessels suspected of being engaged in IUU fishing.[21] This commitment was confirmed in January 2004, when the Prime Minister of Mauritius announced that inspections will be conducted to ensure vessels without 'proper licensing, without precise CDS papers or without effective flag-State control, would be refused access to unloading and support facilities at Mauritius.'[22] In October 2004 Mauritius joined *CCAMLR* as an acceding state.

Namibia attended the *CCAMLR* Commission meeting in 1998 and subsequently became a member of the Commission in 2001.[23] The Namibian official attending the 2001 *CCAMLR* meeting informed the Commission that the Namibian government was reviewing its *Sea Fisheries Ac*t in order to address problems raised by Namibian ports being used to land Toothfish caught within the Convention Area in contravention of *CCAMLR* or coastal state measures.[24] In 2000, the *CCAMLR* Standing Committee on Observation and Inspection congratulated Namibia in effectively shutting down Walvis Bay as a port for IUU fishing activity.[25]

The Seychelles, another known port State and non-member of *CCAMLR*, has recently closed its ports to landings of Patagonian Toothfish by vessels flagged to non-members of *CCAMLR*.[26] The Seychelles policy change is arguably attributable to the considerable Swordfish population within the State's EEZ. In seeking to develop this fishery, the Seychelles will require the cooperation of the international community and openly undermining *CCAMLR* efforts could possibly hinder this process.[27] The Seychelles demonstrated further support for responsible high seas fishing through its ratification of the *UN Fish Stocks Agreement* on 20 March 1997.

There have been several instances of port States closing their ports[28] and with reference to *CCAMLR* waters, coastal States have on occasions encouraged port States to exercise their jurisdiction over IUU fishing vessels. In 2000 the

---

20 Statement by Acting Prime Minister on Illegal Fishing, Mauritian Parliament, 31 October 2000 (copy on file)

21 Ibid.

22 I. Macdonald, Minister for Fisheries, Forestry and Conservation, 'Mauritius vows to close its port to illegal fishers 27 January 2004 <www.mffc.gov.au/releases/2004> (29 January 2004).

23 *CCAMLR*-XVII (1998) Report of the Seventeenth Meeting of the Commission, paragraph 1.7

24 Ibid, paragraph 2.2.0.

25 *CCAMLR*-XIX (2000) Annex 5, Report of the Standing Committee on Observation and Inspection, paragraph 2.56.

26 Ibid.

27 Ibid.

Polar Viking, a New Zealand flagged vessel, was turned away from the port of Montevideo, Uruguay, on the ground that it had engaged in IUU fishing.[29] The same year Australia refused entry to two Belize flagged fishing vessels. Both the *Cisne Rojo* and *Cisne Azul* were denied entry to Australian ports on the grounds that they failed to show that a catch of Patagonian Toothfish landed previously in Mauritius had been caught in accordance with *CCAMLR* conservation measures.[30] During the 2003 *CCAMLR* meeting, the Namibian delegate reported that the *Virgin of Carmen* had been denied permission to land Toothfish following an inspection pursuant to the CDS conservation measure.[31] In 2004 South African enforcement officers alerted authorities in Singapore to prevent the sale of Toothfish suspected of being harvested illegally from South African waters.[32]

Shortly after the introduction of the CDS increasing amounts of Toothfish were documented as having been harvested in areas beyond the *CCAMLR* area of application.[33] Scientific evidence suggests that it is unlikely these quantities could in fact have been harvested from the areas claimed by fishers.[34] This development suggests that the CDS is having an impact on IUU operators. Faced with port States demanding completed CDS documents, IUU operators have sought to evade the application of the CDS by alleging the catches were taken outside the CCAMLR area of application. The challenge for *CCAMLR* contracting parties is to work towards a solution which eliminates the opportunity for fraudulent reporting of catch location.

6.2.4 Verification of CDS Documentation

Under the CDS, contracting parties are required to ensure each landing at their ports, or transhipment to their flagged vessels, is accompanied by a completed CDS document. Whilst the document must detail the IMO or Lloyd's registration number of the vessel and provide details such as the dates the catch was taken and the *CCAMLR* statistical sub-area or division the Toothfish was harvested from, (or FAO

---

28 For a discussion on coastal State legislation see Louise de La Fayette, 'Access to Ports in International law' (1996) 11(1) *International Journal of Marine and Coastal Law* 1, 6-9.

29 ISOFISH, *Toothfish test boat turned away* 28 June 2000. The *Polar Viking* was registered with New Zealand as a sea going vessel only.

30 Greenpeace, *Australia leads on pirate port battle* 6 April 2000.

31 *CCAMLR*-XXII (2003) Report of Twenty Second Meeting of the Commission, paragraph 6.30 and Annex 5, paragraph 3.18-3.19. The vessel was flagged to the Netherlands Antilles. Landing was denied on the ground that the vessel did not possess either a fishing licence, CDS, VMS nor have a scientific observer on board.

32 J.Yeld, *The Cape Argus*, 'SA leads Swoop on Suspect Toothfish vessel' 18 October 2004. <www.allafrica.com/stories/printable/200410180685.html> (20 October 2004)

33 *CCAMLR*-XXI (2002) Report of the Twenty- First Meeting of the Commission, paragraph 7.8.

34 Ibid, paragraphs 8.2-8.4. This based on the belief that the quantities attributed to the areas outside the Convention Area could not be harvested from those areas.

statistical area if harvested outside the Convention Area)[35] there is no mechanism for verifying the accuracy or authenticity of the information provided. This means that the scope for high level fraud through the entry of false dates and/or fishing location details is considerable. Even in the absence of fraud, several port States lack the expertise or resources to rigorously inspect each CDS document or to identify the accompanying catch of Toothfish.[36]

The *CCAMLR* Commission has addressed the above deficiency in part through Resolution 16-XIX, which provides that flag States participating in the CDS may, on a voluntary basis, take steps to ensure that vessels authorised to fish or tranship Toothfish, maintain an operational VMS. A satellite based VMS would enable the flag State to monitor a vessels' location and activity. Such information could be used to verify the accuracy of the documentation submitted under the CDS.

If implemented uniformly and accurately, the VMS can become an integral component in the exercise of flag State responsibility in the context of high seas fisheries. It does, however, remain dependant upon flag State implementation for its success. The credibility of the VMS can be easily undermined by a lack of contracting party support or the submission of inaccurate data. Whilst non-contracting parties are not obligated to require flag vessels install a VMS (maintain it in an operational state and submit accurate data) to support CDS information, to do so would assist in rebutting the presumption of non-compliance raised by conservation measures 10-07.

On a more general level, *CCAMLR* member States agreed to implement the VMS from March 1999,[37] and further agreed to provide VMS information on the movements of their flagged vessels when entering and leaving the Convention Area (and crossing between *CCAMLR* statistical areas). However verification of the accuracy of that data remains a difficult issue. For example, when the *Viarsa* was arrested in 2003 by Australian authorities, the actual location of the vessel did not correspond with the VMS data being provided by the vessel to Uruguay, the flag State.[38]

In part, an increase in fraudulent VMS data might be attributable to the implementation of the CDS. As stated, soon after the CDS took effect, IUU vessels started reporting that the Toothfish onboard had been harvested in locations outside the Convention Area.[39] It is very unlikely the catches attributed to areas outside

---

35 The *CCAMLR* areas adopt the FAO numbering system, the distinction being made here is between the Convention Area and oceans beyond that area.

36 The two last statements are supported by findings in *CCAMLR*-XXIII/BG/27 Rev.2 *Illegal Toothfish Trade: Introducing Illegal Catches into the Markets,* document submitted by ASOC. (copy obtained with the permission of the ASOC, copy on file).

37 CM 10-04 (2002).

38 D.Fickling, 'Toothfish poachers arrested after 7000km chase' *Manchester Guardian Weekly*, 10 September 2003. (copy on file).

the Convention Area have been accurately reported. It is more likely to be the case that IUU catches have been sold under the CDS through falsified documentation supported by inaccurate VMS data.[40]

Whilst Resolution 16-XIX does not oblige parties to supply any VMS data collected to support the CDS, the Commission has indicated its preference for such an obligation through an optional clause in the CDS conservation measure.[41] However, as discussed, to be effective, accurate VMS data must be provided by the flag State. One option for authenticating the CDS documentation, albeit again applied on a voluntary basis, is contemplated in the CDS conservation measure. Contracting parties or cooperating non-contracting parties may require additional verification of CDS documents from 'flag States by using, *inter alia*, VMS in respect of catches taken on the high seas outside the Convention Area, when landed at, imported or exported from its territory.' This would allow verification that Toothfish claimed to have been caught outside the Convention Area have in fact been caught there. This option is very similar to the proposal in Resolution XIX and shares the same weakness- it is dependent upon accurate VMS data being provided by the flag State which, on available evidence, is not guaranteed.

A recent development shows more promise of addressing the problems with fraud in the documents submitted under the CDS and the inaccuracies in supporting VMS data. The trial of a centralised and electronic VMS (C-VMS) whereby data would be provided to the flag State and the *CCAMLR* secretariat in 'real time' was agreed to at the 2003 *CCAMLR* meeting.[42] Five contracting parties participated in the trial. The opportunities for misreporting a vessel's position are considerably reduced if the *CCAMLR* Secretariat simultaneously receives vessel position data and distributes that data to the 32 *CCAMLR* State parties. The C-VMS indicates the potential of the VMS to enhance the effectiveness of port State controls to eliminate IUU fishing by reducing the opportunities for accessing legitimate markets through fraudulently completed CDS documents.

The Commission adopted the C-VMS in 2004 via the amendment of conservation measure 10-04. Contracting parties are obliged to forward VMS reports submitted by their flag vessels 'to the *CCAMLR* Secretariat as soon as possible but (i) not later than four hours after receipt for those exploratory longline fisheries subject to conservation measures adopted at *CCAMLR*-XXIII; or (ii) following departure from the Convention Area for all other fisheries'.[43] The measure applies to the Convention Area only.[44] Flag States may also require their vessels to communicate VMS reports

---

39 Reports of catches from statistical areas 51 and 57 which scientist believe are unable to support the size of catch reported. *CCAMLR*-XXI (2002) Report of the Twenty-First Meeting of the Commission, paragraphs 8.2-8.4.

40 Ibid, See Annex 5, Report of the Standing Committee on Observation and Inspection, paragraphs 5.12-5.29.

41 CM10-05 (2004) paragraph 14.

42 *CCAMLR*-XXIII, Report of the Twenty Third Meeting of the Commission, Annex 5, paragraph 3.31.

to the Secretariat in parallel.[45] The Secretariat is to place the information on a password protected section of the *CCAMLR* website with new information posted daily. Whilst the exact location of individual vessels will not be provided, the area or sub-area, in which the vessel is operating or transiting, will be nominated.[46] Confidentiality requirements apply to VMS data provided but it can be accessed to conduct active surveillance in specified CCAMLR areas, inspections, or to verify CDS documents.[47]

The *CCAMLR* Commission is also considering further enhancing the effectiveness of the CDS by implementing an electronic web-based CDS (E-CDS), was canvassed at the 2003 *CCAMLR* meeting.[48] A trial of the proposed Scheme was conducted in 2003 by a limited number of States, including coastal States, port States and market States.[49] There was wide support for the introduction of electronic reporting, subject to the continuation of the trial for the 2004 intercessional period, with all member States welcome to participate.[50] The advantage of an electronic web-based CDS is that it can be completed in 'real time' which reduces the scope for fraud.[51] When coupled with the C-VMS data, also proposed to be obtained in 'real time', the net will close upon IUU fishers making it increasingly difficult to avoid *CCAMLR* conservation measures. An example of the increasing difficulties IUU fishers will encounter when seeking to unload or sell IUU catches is the announcement at the 2004 *CCAMLR* Commission meeting by the USA delegation that the USA would no longer accept Toothfish not accompanied by an electronically based CDS.[52]

6.2.5 WTO Considerations

The possibility that port State policies which deny permission to land catches harvested in violation of regional conservation and management measures, being WTO/GATT inconsistent, was raised during negotiations at the UN Fish Stocks Conference. It was however concluded that GATT provisions do not apply to direct imports from the high seas.[53] This would not cover market State controls which apply to the trade in all Toothfish and products, including the IUU trade.

Notwithstanding this conclusion, it would be prudent for port State controls to

---

43 CM10-04 (2004), paragraph 11.

44 Ibid, paragraph 7.

45 Ibid, paragraph 12.

46 Ibid, paragraph 19.

47 Ibid, paragraphs 18 and 22.

48 *CCAMLR*-XXII (2003) Report of the Twenty Second Meeting of the Commission, paragraph 7.7.

49 Ibid, Australia, Chile, South Africa, UK, Spain, Japan and USA were involved in the 2003 trial.

50 Ibid, 7.8. The trial was endorsed at *CCAMLR*-XXIII (2004) Report of the Twenty Third Meeting of the Commission, Annex 5, paragraph 4.17.

51 Currently contracting States can inspect CDS documents via a secure link on the *CCAMLR* website.

52 *CCAMLR*-XXIII (2004) Report of the Twenty Third Meeting of the Commission, paragraph 7.2.

be WTO consistent. The right of port States to close ports to fishing vessels was brought to international attention during the *Case Concerning the Conservation and Sustainable Exploitation of Swordfish Stocks in the South-Eastern Pacific Community*.[54] State practice indicates that from a fisheries viewpoint there is no support for a system of open ports.[55] Norway has closed its ports to high seas catches taken without a quota and Chile has closed its ports to Spanish vessels.[56]

The international inclination towards trade liberalisation[57] was confirmed during the Uruguay Round negotiations[58] which culminated in the inception of the WTO in 1995.[59] However it is recognised that trade barriers are required at times to prevent the spread of disease, protect consumers and the environment. A decision to establish the Committee on Trade and the Environment was made during the final Uruguay Round negotiations at Marrekesh. A number of multilateral environmental agreements already contain restrictions on trade. For example CITES[60] bans the trade in an agreed list of endangered species.

The principle defence to the argument that the CDS amounts to a discriminatory trade measure and as such is contrary to Article XI of *GATT*, would be that the Scheme is a permitted exception under Article XX, the full text of which is reproduced below.

---

53 Earth Negotiations Bulletin, Summary of UN Fish Stocks Conference, Vol. 7(15). Article XI of GATT states that no restrictions shall be instituted or maintained by any contracting Party on the importation of any product of the territory of any other contracting party. However it is not arguable that fish caught on the high seas are the product of the territory of the flag State of the vessel. This point is also made by A. Serdy in 'See you in Port: Australia and New Zealand as Third Parties in the Dispute between Chile and the European Community over Chile's denial pf port access to Spanish vessels fishing for Swordfish on the high seas' (2002) 3 *Melbourne Journal of International Law* 79,83. See also 85-86 for a discussion of Article V relating to the freedom of transit through the territory of contracting Parties.

54 *Chile v. European Community* ITLOS, Case No. 7.

55 See F.O.Vicuna, *The Changing Law of High Seas Fisheries* (1999) 261-6.

56 O.S.Stokke, 'Managing Fisheries in the Barents Sea Loophole: Interplay with the UN Fish Stocks Agreement' (2001) 32 *Ocean Development and International Law* 241, 245. L.De La Fayette, 'Access to Ports in International Law' (1996) 11 *International Journal of Marine and Coastal Law* 1, 9. Serdy, 'See you in port' above, 80, 94.

57 This is alluded to in the *Code of Conduct*, Article 11.2.5. The Uruguay Round of trade negotiations

58 1986-1994.

59 Prior to 1995 the *General Agreement on Trade and Tariffs* (GATT) 1948, guided international trade.

60 Opened for signature 3 March 1973, 12 ILM 1085 (entered into force 1 July 1975)

GATT 1994, Article XX

Subject to the requirement that such measures are not applied in a manner which would constitute a means of arbitrary or unjustifiable discrimination between countries where the same conditions prevail or a disguised restriction on international trade, nothing in the present Agreement shall be construed to prevent the adoption or enforcement by any contracting party of measures:

(a) necessary to protect human, animal or plant life or health;

(g) relating to the conservation of exhaustible natural resources if such measures are made effective in conjunction with restrictions on domestic production or consumption.

The application of *GATT* and Article XX was discussed by the WTO Panel and Appellate Body in the *Shrimp/Turtle* dispute.[61] The dispute centred on the introduction of sea turtle protection measures, by the USA, which aimed to reduce the high levels of turtle mortality associated with the shrimp fishery.[62] The USA sought to impose trade restrictions on all shrimp imports not otherwise certified by the Department of State as being caught in accordance with sea turtle protection measures. The trade restrictions were objected to by a number of Asian States and the matter was presented to the WTO Panel. The Panel acknowledged that under WTO rules there are rights to protect the environment but that the import ban imposed under USA domestic law was not consistent with Article XI(1) and furthermore, was not justified under any exceptions in Article XX.

The USA appealed to the WTO Appellate Body who agreed that the import restrictions on shrimp did fall within Article XX(g). However, the Appellate Body concluded that the measure was administered in an arbitrary and discriminatory manner, contrary to the head paragraph (or chapeau) of Article XX.[63]

[61] WTO *Appellate Body Report on United States import prohibition of certain Shrimp and Shrimp Products,* 38 ILM 118. (12 October 1998)

[62] K.Larson, 'Fishing for a Compatible Solution: Toothfish Conservation and the World Trade Organisation' (2000) 7 *The Environmental Lawyer* 123, 143.

[63] A summary of *the Shrimp Turtle case* can be found in Larson, Ibid, 139-148. The US provided financial assistance to a number of Caribbean countries to assist with the implementation of the turtle protection measures. Such assistance was not provided to the Asian countries complaining of the US measures.

The *CCAMLR* CDS is different in some important respects to the shrimp import restrictions sought to be imposed by the USA. First, the Scheme is the product of a multilateral organisation with 32 member States, whilst the restrictions imposed by the U.S.A. were unilateral and domestic in nature.[64] Second, the Toothfish trade restrictions imposed through the CDS do not apply on the basis of country of origin (whereas the shrimp import restrictions did) but on the basis of completed documents, or rather unsatisfactorily completed documents.[65]

The *CCAMLR* Secretariat has expressed quiet confidence that the CDS is compliant with WTO provisions.[66] The WTO Committee on Trade and Environment has considered the *CCAMLR* CDS, but has not reached consensus on the issue.[67] The WTO Secretariat has, however, indicated that the *CCAMLR* Scheme, 'provides examples of appropriate and WTO-consistent (i.e. non-discriminatory) use of trade measures in multilateral environmental agreements.'[68]

6.2.6 Market Controls

Market State controls over the landing, import, export and re-export dovetail neatly into port State controls.[69] The CDS seeks to identify (and verify) the origin of the catch, ascertain whether it has been caught in accordance with conservation measures and limit trade in IUU Toothfish by prohibiting the landing or transhipment of Toothfish unless accompanied by satisfactory CDS documents. Information available from CDS data shows the main importers of Toothfish are Japan, the U.S.A., the European Community (particularly Spain) and China and to a lesser degree Canada.[70] China is the only non-contracting State. Thus, good opportunities to enforce port and market State controls through *CCAMLR* members do exist.

---

64 Ibid, 149. The WTO has expressed a preference for environmentally based trade restrictions formulated through multilateral consensus rather than restrictions imposed unilaterally.

65 Ibid, 150.

66 D. Miller, E. Sabourenkov and N. Slicer, 'Unregulated Fishing – The Toothfish Experience, in M. Richardson and D.Vidas (eds) *The Antarctic Treaty System for the Twenty First Century* (2005 16. The authors note that the CDS meets three key elements of WTO concern. That is it is non discriminatory, it is transparent in its multilateral nature and has a clear linkage to a policy which aims to conserve the resources concerned.

67 Ibid. 16. Per WTO Committee on Trade and Environment Report of meeting held 24-25 October 2000 and meeting 12 December 2000. DocumentWT/CTE/M/25.

68 Ibid, WTO, 'Environmental benefits of removing trade restrictions and distortions: The Fisheries sector- Note by the Secretariat' WTO Committee on Trade and Environment, WTO 16 October 2000. (Doc.WT/CTE/W/167).

69 Generally trade documentation or controls apply at the point of landing and accompanies fish or fish products entering the international market. Catch certification applies at the point of harvest and applies to the landing or transhipment of fish. See Miller, et al 'Unregulated Fishing- The Toothfish Experience' above, 16-17.

70 Miller, ibid, 20.

Several *CCAMLR* member States have introduced trade codes to track Toothfish imports and exports. In 1999, the USA, introduced a customs control based upon the Harmonised Shipping Code whereby unique numbers are assigned to commodities and can be used to track their movement internationally.[71] The USA, Canada, Norway and Australia reported the introduction of trade codes effective from early 2000.[72] However, in 2002 the USA reported to the Commission that the CDS still allowed Toothfish to be landed or transhipped under the cover of fraudulent catch documents.[73]

As a significant importer of Toothfish,[74] the USA, proposed a number of changes to its domestic controls to ensure only legally harvested Toothfish and Toothfish products entered its territory. These changes included the classification of Toothfish as either Antarctic or Patagonian and as fillet or non-fillet fish. The requirement to obtain an American dealer permit, which would effectively bring the dealer/agent under the jurisdiction of USA law, was implemented from 2000.[75] In 2003 the World Customs Organisation adopted standard custom tariff codes for Toothfish products which will assist in tracking the movement of Toothfish.[76] These codes are applicable from 1 July 2007 and *CCAMLR* contracting parties are urged to voluntarily adopt the codes as soon as possible.[77]

There are reports of success in deterring IUU fishing from the implementation of trade and port State controls. It has already been observed that Toothfish unaccompanied by appropriate CDS documents has commanded a lower price than Toothfish with satisfactorily completed CDS documents.[78]

---

71 *CCAMLR*-XXIII/BG/27/Rev.2 *Illegal Toothfish Trade,* above, 4.

72 *CCAMLR*-XVIII (1999) Report of the Eighteenth Meeting of the Commission, paragraphs 2.20-2.24. The USA has in fact implemented trade codes since 1998 and the 2000 codes reflect further refinement.

73 *CCAMLR* –XXI (2002) Report of the Commission, paragraph 8.60.

74 In 2000 USA imports represented about 61% of the legal *CCAMLR* catch and in 2003 the percentage increased to 85%. *CCAMLR*-XXIII/BG/27/Rev.2, *Illegal Toothfish Trade*, above, 4

75 Ibid.

76 *CCAMLR*-XXIII (2004) Report of the Standing Committee on Observation and Inspection, Annex 5, paragraph 4.10.

77 *CCAMLR*-XXIII (2004) Report of the Twenty Third Meeting of the Commission, paragraph 7.4.

78 *CCAMLR*-XIX (2000) Report of the Standing Committee on Observation and Inspection , Annex 5, paragraph 2.3.

6.3 *CCAMLR* IUU Vessel Lists

6.3.1 Legal basis

Article VIII of the *Compliance Agreement* requires State parties to 'exchange information amongst themselves, either directly or through the FAO, with respect to activities of fishing vessels flying the flags of non-parties that undermine the effectiveness of international conservation and management measures.' The *IPOA-IUU* urges States to act through relevant RFMOs to establish records on those vessels engaged in or supporting IUU fishing and to cooperate in the exchange of information on such vessels.[79] The *UN Fish Stocks Agreement* also emphasises the role of RFMOs by requiring States which are members of a relevant RFMO to 'exchange information with respect to the activities of fishing vessels flying the flags of States which are neither members of the organisation nor participants in the arrangement and which are engaged in fishing operations for the relevant stocks.'[80]

Gathering and collating information collected about IUU fishing vessels into databases is not a *CCAMLR* initiative. In 1999, *ICCAT* established a listing of IUU vessels pursuant to a Resolution calling upon contracting parties (as well as cooperating non-contracting parties, entities or fishing entities) which import frozen tuna and tuna like products to submit detailed information including the name of the vessel which caught and produced the fish, the flag state, the species of tuna caught, the area the catch was taken, vessel registration and details of the vessel owner.[81] In 2002 *ICCAT* resolved to establish a list of non-contracting party fishing vessels presumed to have carried out IUU fishing within the convention area.[82] The presumption is premised upon the production of evidence that vessels, *inter alia,* harvest tuna within the Convention area without being registered to do so, fail to report catches to *ICCAT* or make false reports, fish during closed seasons or tranship catches with vessels on the IUU list.[83]

---

79 *IPOA-IUU*, Paragraph 80.4 & 80.5.

80 *UN Fish Stocks Agreement*, Article 17(4). Article 17 applies to RFMO or regional fisheries management arrangements.

81 *ICCAT* Resolution 98-18. 'Unreported and Unregulated Tuna Catches by large-scale Longline Vessels in Convention Area' Entered into force 21 June 1999. Replaced by *ICCAT* Resolution 03-15 'Resolution by ICCAT concerning Trade Measures'.

82 *ICCAT* Resolution 02-23 'Recommendation by ICCAT to establish a list of vessels presumed to have carried out Illegal, Unreported and Unregulated Fishing activities in the ICCAT Convention Area'

83 Ibid, Paragraph 1. This listing is not exhaustive.

84 *CCAMLR*-XX (2002) Report of the Twentieth Meeting of the Commission, paragraph 5.19.

6.3.2 The Establishment of the *CCAMLR* IUU Lists

The establishment of a *CCAMLR* IUU Vessel List stems from the 2001 meeting, during which the Commission accepted advice from the Standing Committee on Observation and Inspection on the issue and determined that the Secretariat would compile a list of flag of convenience States.[84] The Commission agreed that two IUU Vessel Lists be established, one for non-contracting parties whose vessels are engaged in IUU fishing and a second list detailing member States' vessels engaged in IUU fishing.[85]

Lengthy discussions took place during the 2003 *CCAMLR* meeting about the criteria for entry upon, and removal from, the IUU Lists.[86] The Russian Federation expressed an opinion that:

> Conservation measure 10-06 violates the balance of the rights and duties of a Flag State. ... to include a vessel in the proposed IUU Draft List, purely on the grounds of any information about any alleged violations of this conservation measure. ...the burden of proving this particular vessel did not participate in IUU fisheries... is on the flag State.[87]

Spain also raised concerns about undesirable impact of IUU Vessel Lists. The Spanish delegate observed that:

> It appears that this measure is pushing Contracting Parties to rapidly deregister their IUU vessels. As a consequence, IUU vessels and companies associated with them move to operate under the flags of non-compliance, otherwise known as flags of convenience or open registries. ...In doing so we export the problem outside the Commission but the devastating action of these IUU Vessels and companies still affect the Southern Ocean.[88]

The problem raised by Spain that of relocating the IUU fleet, is real. As discussed in Chapter 3, IUU operators are increasingly highly organised and mobile. The challenge for RFMOs and national governments is to reduce the opportunities for the IUU operators to relocate. That is, reduce the number of ports which will accept fish without accompanying CDS documents, increase the number of States which apply market controls and gradually envelop the seas with a network of cooperation so that the IUU operator and fisher have nowhere to retreat to conduct their business. It is only by making the environment in which those involved in the IUU industry

---

85 CM 10-07 (2003) and 10-06 (2004).

86 *CCAMLR*-XXII (2003) Report of the Twenty Second meeting of the Commission, paragraphs 8.19-8.67.

87 Ibid, Paragraph 4.44.

88 Ibid, Paragraph 8.42.

89 CM 10-07 (2003), paragraphs 3-9. This is based on the presumption that the vessel is undermining

operate less attractive, that the economic pendulum may swing in favour of legal fisheries.

6.3.3 Criteria and Procedure for listing vessels

The procedure for, first, including vessels on the proposed IUU Vessel List or the annual IUU Vessel List and second, considering their subsequent removal, is outlined in the applicable *CCAMLR* conservation measures. Currently, a non-contracting party vessel qualifies for reporting to the Commission for review by the Standing Committee on Implementation and Compliance if it is sighted engaging in fishing activities within the Convention Area or is denied port access, landing or transhipment. The Standing Committee on Implementation and Compliance then makes recommendations to the Commission to add, delete or retain the reported vessels from the proposed List.[89]

The procedure for *CCAMLR* contracting parties is more complex given the rights of parties to comment on inclusions on the Draft IUU Vessel List.[90] Given the opportunities for contracting parties to present evidence to the Commission indicating that their identified flag vessels have not engaged in fishing activities in contravention of conservation measures, the comments, cited above, of the Russian delegation at the 2003 *CCAMLR* meeting, seem unfounded. Rather than imposing the burden of proof of innocence upon the flag State, the measure facilitates the provision of information about the activities of flag vessels to clarify questions raised about their fishing activities. The measure states unambiguously that 'in order to establish the IUU Vessel List, evidence… shall be required that vessels flying the flags of the contracting party concerned have' engaged one of a list of seven activities including: fishing without a licence; fishing during a closed season; failing to record or declare catches; or transhipping fish with vessels already identified for the IUU Vessel List.[91]

The proposed list of non-contracting party vessels for 2004 contained twelve vessels flagged to known flag of convenience States including Togo, Honduras and Equatorial Guinea.[92] The three contracting party vessels listed were flagged to Uruguay.[93] Whilst a number of refinements to the listing criteria would enhance

the effectiveness of *CCAMLR* conservation measures.

90 CM10-06 (2004) paragraphs 1-4, 7-14.

91 Ibid, paragraph 5. This listing of criteria is similar to the *ICCAT* list.

92 *CCAMLR*-XXIII (2004) Report of the Twenty Third Meeting of the Commission, Appendix 3 to Annex V.

93 Ibid.

94 For example inserting definitions of the terms 'fishing' and 'transhipment' as recommended by the

the transparency of the IUU Vessel List process,[94] the foundations have been laid for an effective deterrence strategy. Contracting parties whose vessels are included on the IUU Vessel List are required to address the IUU fishing activity which may necessitate withdrawing vessel registration or licences, the denial of the right to land or tranship catches and the banning of imports from such vessels.[95] Non-contracting parties whose flag vessels are identified under conservation measure 10-07 (2003) are requested to take similar action against their vessels.[96]

As evidenced in the discussions at the 2003 and 2004 *CCAMLR* Commission meetings, no State likes to have their flag associated with IUU fishing. It is akin to the label 'flag of no responsibility' suggested by Stokke and Vidas.[97] Herein lies the strength of the IUU Vessel Lists. They are applied through regional and international cooperation, do not require the consent of flag States and contribute to making the IUU operating environment uncomfortable and unattractive.

6.3.4 Australian proposals for improvements to the IUU Vessel List

In 2004 the Australian delegation proposed two primary amendments to the conservation measures under which the IUU Vessel lists have been established. The amendments are aimed at improving the usefulness of the IUU vessel lists which are 'currently limited by their restricted availability and the speed at which the information on the vessels on the lists becomes outdated.'[98] The two improvements were to:

- Make the lists publicly available, and
- Have the information updated on a regular basis with updates to be made as soon as information on listed vessels changes.

This would be in accordance with the *IPOA-IUU* which calls upon all States to 'publicise widely, including through cooperation with other States, full details

SCIC. *CCAMLR*-XXIII (2004) Report of the Twenty Third Meeting of the Commission, Annex 5, paragraph 2.13.

95 CM10-06 (2004) paragraph 18.

96 CM10-07 (2003) paragraph 13.

97 O. Schram Stokke and D. Vidas, 'Regulating IUU Fishing or Combating IUU Operations' paper prepared for OECD *Workshop on Illegal, Unreported and Unregulated (IUU) Fishing Activities* Paris, 19-20 April 2004. <http://www.oecd.org/documents/5/0.2340.en_2649_33901_21007109_1_1_1_1.00.html> The paper has also been published in OECD, Fish Piracy: Combating Illegal, Unreported and Unregulated Fishing (2005).

98 *CCAMLR*-XXIII/48 Improvements to the *CCAMLR* IUU Vessel Lists, 12 October 2004. Submission of the Delegation of Australia, 1. Copy provided to author with permission of the delegation. (Copy on file).

99 *IPOA-IUU*, paragraph 32.

of IUU fishing and actions taken to eliminate it.'[99] The *IPOA-IUU* does however; place a caveat upon disclosure that it be consistent with applicable confidentiality requirements.'[100] The basis for the Australian proposal on public availability is that access to the *CCAMLR* IUU Vessel List is currently restricted to *CCAMLR* contracting parties. Thus, non-contracting parties wishing to assist in deterring IUU fishing through the application of the CDS, by either limiting port access or refusing permission to land catches from IUU vessels, are hampered by their inability to access the *CCAMLR* IUU vessel list.[101]

Furthermore IUU vessels can and do change names and flags quickly and updating the list yearly (as is currently the case) through the *CCAMLR* Commission meetings means the List can become redundant within a short time.[102] Provision for the updating of information as it becomes available will assist RFMOs and coastal States in keeping up with IUU operators.

It is the link to timely and accurate information which is crucial. *ICCAT* places IUU vessel lists on the *ICCAT* website and provides the list to other RFMOs.[103] The FAO has established the HSVAR and has been encouraged to develop a central repository for information relating to IUU fishing.[104]

Following consideration of the Australian submission, conservation measure 10-06 was amended by the Commission with the insertion of a number of new paragraphs. The more significant amendments are:

- New paragraph 3, whereby contracting parties are obliged to submit information that vessels flying the flag of another contracting party have engaged in activities set out in paragraph 5 of the measure (and reproduced above). Such information is to be provided within 30 days of having become aware of it.
- New paragraph 6, detailing the information to be included in IUU Vessel Lists (see below for full version of paragraph 6).
- The amendment of the 2002 version paragraph 8 to replace the word 'may' with 'shall', so that contracting parties are now obliged to submit to the Executive Secretary any additional information within 30 days of having become aware of such information and at the latest 30 days before the start of the *CCAMLR* meeting.

---

100 Ibid.

101 Ibid.

102 Ibid.

103 Ibid. Other RFMOs include the *Indian Ocean Tuna Commission* and the *Inter-American Tropical Tuna Commission.*

104 See the FAO Technical Consultation to review progress and promote the full implementation of the International Plan of Action to Prevent, Deter and Eliminate IUU Fishing and the International Plan of Action for the management of Fish Capacity, Rome 24 - 29 June 2004, FAO Fisheries Report No. 753, Paragraph 73.

- A new paragraph 11, which instructs the Executive Secretary to invite non-contracting parties cooperating with the Commission and CDS to submit any evidence or documented information regarding vessels on the Provisional List or IUU Vessel list. To this end, a new paragraph 25 instructs the Executive Secretary to circulate to cooperating non-contracting parties a copy of the Provisional IUU List and the IUU Vessel List. It is noted that non-contracting parties are not given the opportunity to initiate a vessel listing.
- A new paragraph 19, which instructs the Executive Secretary to place the IUU Vessel List on the *CCAMLR* website and provide a copy of the list to the FAO and appropriate RFMOs to enhance cooperation between organisation and to prevent, deter and eliminate IUU fishing.
- A new paragraph 20, obliging contracting parties obtaining new or changed information in relation to the details required in paragraph 6, to notify the Executive Secretary, who shall then provide *CCAMLR* contracting parties an opportunity to comment by placing the new information on the secure section of the *CCAMLR* website for seven days.

6.4 *CCAMLR* Licensed Vessel List

Conservation Measure 10-02 was originally adopted by the *CCAMLR* Commission to reiterate the obligation of all contracting parties to exercise their duties as flag States. During the 2004 meeting the measure was amended to require contracting parties to provide detailed information about licensed vessels with the view that such information will be placed on the *CCAMLR* website in a list of licensed vessels. The practical impact of such a list is that *CCAMLR* members and coastal States with Southern Ocean EEZs can access the list of licensed vessels and use that information when conducting patrols (either under the *CCAMLR* system of Observation and Inspection or when exercising coastal State rights within the EEZ).

6.5 *CCAMLR* Resolutions

A number of Resolutions have also been passed by the Commission, aimed at supporting conservation measures or indicating areas where further action might be taken to deter IUU fishing. The subject matter of many Resolutions can be linked to the content of the international instruments reviewed in Chapter 4. The Resolutions relevant to deterring IUU fishing within the Convention Area are listed in Table 6.1 below. Following that, the individual Resolutions are examined in terms of their relevance to deterring IUU fishing.

Resolution Number	Resolution Title and Description
10/XII	Resolution on harvesting of Stocks occurring both within and outside the Convention Area.
14/XIX	Catch Documentation Scheme: Implementation by Acceding States and Non-Contracting Parties
15/XIX	Use of Ports not Implementing the Catch Documentation Scheme for *Dissostichus* spp.
16/XIX	Application of Vessel Monitoring System in the Catch Documentation Scheme.
17/XX	Use of Vessel Monitoring System and other Measures for verification of the Catch Documentation Scheme catch data for areas outside the Convention Area, in particular in FAO Statistical area 51.
18/XXI	Harvesting of *Dissostichus eleginoides* in the areas outside of coastal State jurisdiction adjacent to the *CCAMLR* area in FAO Statistical areas 51 and 57.
19/XXI	Flags of Non-Compliance

Table 6.1 Resolutions supporting conservation measures to reduce IUU Fishing in the *CCAMLR* Convention Area

6.5.1 Regional Cooperation

Resolution 10/XII urges contracting parties to ensure flag vessels, engaging in fishing in marine areas adjacent to the *CCAMLR* Convention Area, do so responsibly and with due respect for the adopted conservation measures. The Resolution applies to all Antarctic marine species. This Resolution reaffirms the obligation under Article XI of the Convention to cooperate with fisheries organisations that exercise jurisdiction over adjacent high seas areas. It is also supportive of the international consensus that increased cooperation between States is required to improve the sustainable management of high seas fisheries and to deter IUU fishing.[105] Finally, the resolution supports the role of the *CCAMLR* Commission as a RFMO.

Resolution 14/XIX notes the evidence that several acceding States and non-contracting parties are not implementing the CDS and urges such States to do so. It further urges member States to make appropriate representations to such States in relation to implementation. As a result of diplomatic approaches, Canada, a contracting party but not a *CCAMLR* Commission member, advised the Commission at the 2003 meeting that it would be able to implement the Scheme.[106]

[105] See generally Part 3.

Resolutions 17/XX and 18/XXI relate to the harvesting of Toothfish outside the Convention Area. In particular, under Resolution 17/XX, all States participating in the CDS, are urged to ensure documents relating to landings or imports are checked by the flag States to verify the information in the documents is consistent with reported VMS data. With the implementation of a dual or centralised VMS, the cross-checking recommended by these Resolutions can be completed via verification with the Secretariat and or the flag State.

6.5.2 Flag State Responsibility and Vessel Authorisation

Resolution 15/XIX urges contracting parties - in the event they are unable to provide an authorised flag State official to monitor landings - to discourage their flag vessels authorised to fish for Toothfish within the Convention Area from using ports of acceding States and non-contracting parties which are not implementing the CDS. To this end, contracting parties are to attach to every fishing authorisation issued to each flag vessel, a list of all acceding States and non-contracting parties that are implementing the CDS.

As mentioned above in the context of the CDS, Resolution 16/XIX encourages States voluntarily, and subject to relevant domestic laws and regulations, to take action to ensure all flag vessels authorised to fish for, or tranship Toothfish on the high seas, maintain an operational VMS.

In 2002, the Commission, concerned at the lack of control by flag States over their vessels on the high seas, passed Resolution 19/XXI. Whilst the primary responsibility of the flag State for its own vessels is expressly acknowledged, contracting parties together with non-contracting parties are urged to take measures to ensure to the greatest extent possible, that their own nationals do not support or engage in IUU fishing and do not work on flag of convenience vessels within the Convention Area.

6.6 Conclusion

This Part has examined the management of Southern Ocean fisheries under *CCAMLR* and considered in detail the conservation measures and resolutions adopted to deter IUU fishing both within the *CCAMLR* area of application and in adjacent areas in the absence of a competent RFMO. In the wake of sustained IUU fishing for the Patagonian Toothfish, the *CCAMLR* Commission has been criticised for not acting decisively enough to deter IUU fishing.[107] However, one factor hampering *CCAMLR* success is the fact that fishing vessels flagged to *CCAMLR* contracting

[106] *CCAMLR*-XXII (2003) Report of Twenty-Second Meeting of the Commission, paragraph 7.2.

[107] See for example press releases by prominent NGOs including Greenpeace, ISOFISH, ASOC and COLTO.

parties have been sighted engaging in IUU fishing within the Convention Area, or implicated in IUU fishing through reports of inaccurate VMS data.

The degree to which *CCAMLR* measures will achieve success in deterring IUU fishing (and it is argued that measurable success is already observed) will depend on support from both contracting and non-contracting parties. Measures such as port and market State controls and IUU vessel lists are credible measures which are not reliant on flag State consent to implement. The IUU Vessel List for contracting party vessels is one way in which non-compliant contracting parties can be 'encouraged' to take action to ensure their flagged vessels comply with all *CCAMLR* conservation measures.

In summary, although *CCAMLR* is one of many RFMOs acting to deter IUU fishing on a regional basis, the measures adopted by the Commission do provide a useful working model of regional cooperation. The activities of coastal States with nearby or overlapping EEZs are fundamental to supporting regional fisheries measures and this is acknowledged in the *IPOA-IUU*. The following Part examines the measures adopted by Australia to deter IUU fishing within the Heard and McDonald Island's fishing zone, the manner in which they are drawn from the applicable international law and the degree to which they support the *CCAMLR* approach.

International Measures	**Conservation Measures adopted by CCAMLR Commission or initiatives of Commission**
Port and market State controls	CM 10-05(2004) - CDS
Recording of vessels and exchange of information	CM10-06 (2004) and 10-07(2002) -IUU Vessel lists
Increased role for RFMO	Steps taken to address IUU fishing outside the Convention Area and interaction with nearby RFMOs.
Duty to cooperate	Resolution 10/XII
National Measures	Urges contracting parties and non-contracting parties to implement conservation measures.

Table 6.2 Application of Measures to deter IUU fishing by CCAMLR with cross reference to Part 3, Chapter 4

PART 5

National Measures: Australia

Introduction

This Part examines the measures implemented by the Australian Government in response to the growing threat of IUU fishing to its fisheries resources in the Southern Ocean. Although the actions are by their very nature, unilateral, they can be placed within the context of supporting the regional measures adopted by *CCAMLR* and reviewed in the preceding Part, and the overarching international framework reviewed in Chapter 4. Indeed, as will be demonstrated throughout this Chapter, many legal measures implemented by Australia can be referenced back to the *IPOA-IUU*.

Before examining policy and legislative responses, it is necessary to appreciate the management framework for Commonwealth fisheries generally[1] and the management policy specific to the Heard and McDonald Islands' Australian Fishing Zone (AFZ). The Chapter proceeds to consider the impact of IUU fishing in the Heard and McDonald Islands' AFZ. The Commonwealth legislative structure will then be examined. Australian responses to IUU fishing are considered in Chapter 8.

There is very little published work which examines Australia's policy and legislative response to IUU fishing.[2] Hence, factual research is substantiated by cross checking a number of sources including governmental statements and reports, newspaper reports, NGO publications and other media sources. In particular, much of the detail surrounding the arrests discussed in the Chapter is available only through government statements and media commentary and it has, where necessary, been verified through interviews with government officers.

---

1 Australia's history as a federation of self governing colonies means it is a Commonwealth. Consequently there are marine areas governed by the states and by the Commonwealth. This was confirmed through the Offshore Constitutional Settlement 1979, under which the states have sovereignty to the three mile limit and the Commonwealth has authority to legislate over marine areas claimed under international law from 3 to 200 nautical miles. For more on this, including a discussion of relevant High Court of Australia case law, see, M.White, *Pollution Laws of the Australasian Region* (1994) Section 7.1. For the purposes of this study, the term Commonwealth is used to distinguish between state and federal fisheries. The interchangeable terms when discussing national matters include: federal, Commonwealth and Australian. For consistency the term Commonwealth is used throughout this part as it more accurately indicates the national level at which policy and legislative changes have occurred.

2 See n 2 of Chapter 1 which lists the published work including that of the author.

Chapter 7

The Management of Commonwealth Fisheries

7.1 The Heard and McDonald Islands' Australian Fishing Zone

7.1.1 Management of the Heard and McDonald Islands

The Heard and McDonald Islands are external territories. As such, the Commonwealth Parliament is empowered under section 122 of *The Constitution*[3] to 'make laws for the government of...any territory placed by the Queen under the authority of and accepted by the Commonwealth'. Provisions for the governing of the two sub-Antarctic Islands are contained in the 1953 *Heard and McDonald Islands Act.*

In fact the Islands were initially governed by the United Kingdom. Sir Douglas Mawson visited Heard Island in 1930 as the leader of the British, Australian and New Zealand Expedition.[4] The first Australian sub-Antarctic scientific station was established at Atlas Cove, on Heard Island, in December 1947. This followed the indication by the United Kingdom that it was willing to transfer rights to the Islands to the Commonwealth of Australia.[5] For the ensuing 50 years, Australian management of the Heard and McDonald Islands was relatively uneventful.

In 1987 the Islands, together with their 12 mile territorial sea, were proclaimed a Wilderness Reserve with a number of activities including fishing and mining prohibited.[6] The Wilderness Reserve is closed to all fishing and is managed by the Australian Antarctic Division (AAD). The location of the Wilderness Reserve is shown on Map 7.3. The Islands and the Wilderness Reserve area were included on the National Estate in 1983 on the basis that they represent the only unmodified

---

3 *The Constitution of Australia Act 1900* (UK).

4 P. Carroll, 'The History of Heard and McDonald Islands' (2000) <www.btinternet.com/~sa_sa/heard_island/heard.html> (23 February 2001). See also n.6 below.

5 Q. Hanich, 'Enforcement and Compliance Challenges to the Management of the Heard and McDonald Island Fisheries', Research Report No. 17 (2000) Antarctic Cooperative Research Centre.

6 *Heard and McDonald Islands Environment Protection and Management Ordinance* 1987. Section 14 prohibits fishing.

7 Register of the National Estate, Australian Heritage Database: <http://www.deh.gov.au/cgi-bin/ahdb/search.pl>. (25 June 2001)

example of a sub-Antarctic Island ecosystem in the world.[7] The remains of the first Australian National Antarctic Research Expedition Base at Atlas Cove are also considered to be of historical significance.[8]

The Heard and McDonald Islands were inscribed on the World Heritage List in 1997.[9] The listed outstanding natural universal values are:

- Outstanding examples representing major stages of the earth's history, including the record of life, significant ongoing geological processes in the development of landforms, or significant geomorphic or physiographic features;
- Outstanding examples representing significant on-going ecological and biological processes.[10]

7.1.2 The Heard and McDonald Islands' Fishing Zone

Events since 1997 have overshadowed any natural heritage significance the Islands possess. An AFZ was declared around the Heard and McDonald Islands in 1979 when the Commonwealth of Australia proclaimed a 200 mile fishing zone around the Australian mainland and external territories.[11] The existence of an AFZ adjacent to the Islands as a consequence of the Proclamation did not translate into any management policy specific to the waters or fish therein. It was not until 1997 that a policy was formed.[12]

In 1994, as a reflection of the recently in force *LOSC*, new terminology was introduced for Australian maritime zones and an EEZ was declared in relation to the Australian mainland and all external territories.[13] The linkage between the AFZ and the EEZ is examined in Section 7.2.1 below.

7.1.3 Location of the Heard and McDonald Islands' Fishing Zone

The Heard and McDonald Islands are approximately 4000 kilometres southwest of Perth, the capital city of the state of Western Australia. The Islands' AFZ is not circular in shape for it shares a common boundary with the French declared maritime zone adjacent to the Kerguelen Isles. This shared boundary is shown on

---

8 Ibid.

9 Ibid.

10 Australian Department of the Environment and Heritage. <www.deh.gov.au/heritage/worldheritage/sites/antarct/index.html> (12 May 2005).

11 Commonwealth of Australia Government Gazette No.S189, 26 September 1979.

12 See below Section 1.2 of this Chapter.

13 The *Maritime Legislation Amendment Act 1994* (Cth) amended the *Seas and Submerged Lands Act 1973* (Cth), See sections 10A and 3 of the amended Act.

Map 8.1 as a line running northeast to south west. It is an agreed maritime border with France, which was settled via a bilateral treaty between the two nations, signed on 10 January 1983.[14] The location of both groups of islands can be seen on Map 5.1 The Heard and MacDonald Islands' AFZ falls almost entirely within the *CCAMLR* Convention Area. A very small section of the northern most tip of the fishing zone lies beyond the Antarctic Convergence.[15] The reach of the Antarctic Convergence is depicted on Map 5.1 and the significant overlap between the Heard and McDonald Island's AFZ and the *CCAMLR* area of application is evident.

7.2 The Management of Commonwealth Fisheries

7.2.1 Management Framework

Commonwealth fisheries are managed under the *Fisheries Management Act 1991* (Cth) and *Fisheries Administration Act 1991* (Cth). The authority of the Commonwealth Parliament to legislate for the governance of fisheries within the Heard and McDonald Islands' AFZ and indeed all Commonwealth fisheries is in section 51(x) of *The Constitution.* The Australian Fisheries Management Authority (AFMA), a statutory authority established in February 1992, is responsible for the management of Commonwealth fishery resources.[16]

The *Fisheries Management Act* defines both the AFZ and the EEZ. Whilst Australia has exercised rights under international law to declare an EEZ, and subsequently exercises rights over an EEZ adjacent to Heard Island and McDonald Islands, the term AFZ has been retained for the purposes of Commonwealth fisheries management under the *Fisheries Management Act*. Offences under that Act refer, for example, to the offence of fishing without a licence within the AFZ (rather than fishing without a licence within the EEZ). For the present purposes reference will be made to the AFZ in the context of the management of Commonwealth fisheries. Notwithstanding this fact, it is useful to understand the correlation between the two terms.

The EEZ was formally proclaimed via an amendment to the *Seas and Submerged Lands Act 1973* (Cth).[17] Section 10A of the *Seas and Submerged Lands Act* reads as follows:

---

[14] Agreement on Maritime Delimination between the Government of Australia and the Government of the French Republic. 4 January 1982 (entry into force 10 January 1983) [1983] ATS 3.

[15] The Antarctic Convergence represents the boundary of the *CCAMLR* area of application. *CCAMLR* Article 1(1) and (4).

[16] *Fisheries Administration Act*, 1991 (Cth) Sections 5 and 6.

[17] A new Division 1A of Part II was created and new sections 10A – 1OC inserted by the *Maritime Legislation Amendment Act 1994* (Cth).

> It is declared and enacted that the rights and jurisdiction of Australia in its exclusive economic zone are vested in and exercisable by the Crown in the right of the Commonwealth.

Section 10B of the same Act further states that the EEZ may be declared by the Governor-in-Council not inconsistently with Articles 55 or 57 of the *LOSC*. Article 55 recognises the existence of an EEZ as a 'specific legal regime' affording rights and obligations to coastal and other States in accordance with Part V of the *LOSC*. Article 57 stipulates that the EEZ 'shall not extend beyond 200 nautical miles from the baselines from which the breadth of the territorial sea is measured.' The EEZ is defined in section 3 of the *Seas and Submerged Lands Act* to reflect the *LOSC*. Section 3 states, 'Exclusive economic zone has the same meaning as in Article 55 and 57 of the Convention.'

The Australian declaration of an EEZ in 1994 did not involve a revocation of the AFZ declared 15 years prior.[18] The *Maritime Legislation Amendment Act 1994* (Cth) did however specifically amend the definition of the AFZ as it was defined within *Fisheries Management Act*.[19] For practical purposes, the AFZ and the EEZ now 'mirror' each other. The amended definition of the AFZ in section 4 of the *Fisheries Management Act* is:

> (a) the waters adjacent to Australia within the outer limits of the exclusive economic zone; and
>
> (b) waters adjacent to each external territory within the outer limits of the exclusive economic zone.

The declaration of a 200 mile EEZ adjacent to mainland Australia and her external territories afforded Australia an expanded jurisdiction under international law, over matters such as off-shore installations, living and non-living natural resources, marine scientific research and the protection of the marine environment.[20] The actual management and administration of Commonwealth fisheries, however, remains a matter for domestic or national law.[21]

---

18 Although the *Maritime Legislation Amendment Act*, did delete the definition of the Australian Fishing Zone from both the *Sea Installations Act 1987* (Cth) and the *Whale Protection Act 1980* (Cth), substituting in its place ,the term exclusive economic zone.

19 *Maritime Legislation Amendment Act,* Schedule 1.

20 *LOSC*, Article 56.

21 Although the right to declare an exclusive economic zone is recognised under international law and Part V of the *LOSC* imposes a number of rights and obligations on both coastal states and other states.

7.2.2 Fishery Management Policy for the Heard and McDonald Islands' Fishing Zone

It was not until the mid 1990s that Australian authorities turned their attention to the enforcement and management of potential fisheries in the Heard and McDonald Islands' AFZ.

An Exploratory Fishery Interim Management Policy was formulated in 1997 in response to commercial interest in fisheries within the Heard and McDonald Islands' AFZ. The Interim Policy was replaced by the Heard and McDonald Islands' Fishery Management Plan 2002. Whilst the interim policy authorised two Australian companies to fish on a commercial basis within the Heard and McDonald Islands' AFZ, many more were keenly interested in participating in what was regarded as a highly lucrative developing fishery. Three Australian flagged vessels currently hold statutory fishing rights to engage in fishing within the Heard and McDonald Islands' AFZ.[22]

No foreign fishing vessels are authorised to fish within the Heard and McDonald Island's AFZ.[23] Part 6, Division 5, of the *Fisheries Management Act,* addresses the various offences applicable to foreign fishing vessels. It is relevant to note here that the *Fisheries Management Act* employs the term 'boat' and offences are drafted as such. The word 'boat' is defined in section 4 of the *Fisheries Management Act* to mean 'launch, vessel or floating craft of any description.' Notwithstanding the choice of 'boat' by Commonwealth parliamentary drafters, in the interests of clarity, the term 'vessel' will continue to be used throughout this Part as it has been throughout the entire text.

The total allowable catch for Patagonian Toothfish for the 2004/2005 fishing year[24] in the Heard and McDonald Island's AFZ was 2,787 tonnes.[25] This is down slightly on the total allowable catch for 2003/2004 (2873 tonnes) and 2002/2003 (2,879 tonnes).[26] Under the Heard and McDonald Islands' Fishery Management Plan the determination of the total allowable catch for a fishing year must take into account, amongst other listed matters, the total allowable catch (if any) set by *CCAMLR*.[27]

---

22 Section 12, Heard and McDonald Islands' Fishery Management Plan 2002, Pt 3, Div 3, *Fisheries Management Act* 1991 (Cth).

23 Foreign fishing licences may be granted under section 34 of the *Fisheries Management Act 1991* (Cth).

24 Fishing year is defined in the Heard and McDonald Islands' Fishery Management Plan 2002 as a period of 12 months beginning on 1 December in any year (Section 3).

25 Heard and McDonald Islands' Fishery Management Plan 2002 (HIMIF TAC D3) Total Allowable Catch Determination 2004/2005 Season dated 24 November 2004.

26 Heard and McDonald Islands' Fishery Management Plan 2002 (HIMIF TAC D2) Total Allowable Catch Determination 2003/2004 Season dated 20 November 2003. Heard and McDonald Islands' Fishery Management Plan 2002 (HIMIF TAC D1) Total Allowable Catch Determination 2002/2003 Season dated 20 November 2002.

27 Heard and McDonald Islands' Fishery Management Plan 2002, Section 11(2)(a)(i).

Other fish species are also targeted within the Heard and McDonald Island's AFZ. The Mackerel icefish[28] catch limit is set at 1,864 tonnes for the 2004/2005 fishing year.[29] In addition, a number of species are identified as allowable by catch species. These include Unicorn icefish (150 tonnes); skates and rays (120 tonnes); grey rock cod (80 tonnes) and all other species (50 tonnes per species).[30]

7.3 The Arrival of the IUU Fishing Fleet

In October 1997 the first foreign fishing vessel was arrested within the Heard and McDonald Islands' fishing zone for fishing without authorisation under the *Fisheries Management Act*. In that same year, estimates placed as many as 70 IUU fishing vessels operating in the vicinity of the Heard and McDonald Islands.[31] A total of eight fishing vessels have been arrested at the time of writing. Table 7.1 lists the vessels. Whilst this figure seems very low, (an average of one arrest per year) particularly in comparison to the hundreds of arrests of foreign fishing vessels in Australia's northern waters,[32] one should bear in mind that the average value of the catch on board the vessels arrested in Australia's southern most fishing zone, is in excess of $AU1 million.[33] The 191 tonnes of Toothfish on board the *Maya V* apprehended in January 2004 was sold to public tender for more than AU$2 million.[34]

IUU fishers continue to be active in the Heard and McDonald Islands' fishing zone. It has been estimated that IUU fishing in the zone represented around 22%

---

28 *Champsocephalus gunnari.*

29 Heard and McDonald Islands' Fishery Management Plan 2002.

30 Ibid.

31 G. Lugten, 'On the Rise and Fall of the Patagonian Toothfish- Food for Thought' (1997) 27 (5) *Environmental Policy and Law* 401 and 'Fishing Piracy around Antarctica shadows Treaty Meeting' AFP (Christchurch, New Zealand). 12 May 1997. (copy on file)

32 144 vessels were apprehended in 2002. Australian Fisheries Management Authority Annual Report 2002-2003, XI. In 2003, 138 vessels were arrested in the northern AFZ. See, I. Macdonald, Australian Minister for Fisheries, Forestry and Conservation, 'First illegal fishers caught for 2004' 7 January 2004. <http://www.mffc.gov.au/releases/2004/04001m.html> (10 January 2005). On 12 June 2005, the 100th arrest in Australia's northern waters (from the top end off the coast of Queensland, Western Australia and the Northern Territory) was reported. Increasingly, the fishers apprehended in the north are 'no longer traditional village fishermen and are often part of well-organised groups.' I. Macdonald, Australian Minister for Fisheries, Forestry and Conservation, 'Century up in fight against illegal fishing' 12 June 2005 and 'Ocean the new hotspot for criminals' 13 May 2005 <www.mffc.gov.au/teleases/2005/05111m.html> (15 June 2005)

33 This figure is obtained from averaging out the estimated value of the catch on board the eight arrested vessels. It should be noted that the first two arrested vessels in 1997 had small quantities of fish on board valued between $178,000 and $250,000 so they have not been used for averaging purposes.

34 I. Macdonald, Minister for Fisheries, Forestry and Conservation and C. Ellison, Minister for Justice and Customs, 'Toothfish tender nets over $2M', 23 April 2004 <www.mffc.gov.au/releases/2004> (25 April 2004).

of the total legal allowable catch set by *CCAMLR* (and overseen by Australia) for 2004.[35] In providing this estimate to *CCAMLR* the Australian delegation noted that 'estimating the level of IUU fishing activity continues to be difficult because of the limited information available, the remoteness of the region and the evasive behaviour of the IUU fishers.'[36]

Surveillance of the waters, the enforcement of management plans and deterrence of illegal fishers is therefore a priority for the AFMA.[37]

Vessel Name	**Flag State**	**Date of arrest**
Salvora	Belize	16 October 1997
Aliza Glacial	Panama	17 October 1997
Big Star	Seychelles	21 February 1998
South Tomi	Togo	12 April 2001
Lena	Russia	6 February 2002
Volga	Russia	7 February 2002
Viarsa	Uruguay	28 August 2003
Maya V	Uruguay	23 January 2004

Table 7.1 Australian arrests of foreign fishing vessels within the Heard and McDonald Islands' Fishing Zone[38]

35 *CCAMLR*-XXIII BG/24 *Illegal, Unreported and Unregulated Patagonian Toothfish Catch Estimate for the Australian EEZ around the Heard and McDonald Islands,* submitted by the Delegation of Australia. 1 October 2004 (Copy on file, accessed with permission). The Australian estimate comprises whole weight tonnes and includes the 200 tonnes of processed fish (equating to 334 tonnes whole weight) on board the *Maya V* when she was arrested on 23 January 2004.

36 Ibid.

37 AFMA Annual Report 2002-2003. The Australian Government Department of Agriculture, Fisheries and Forestry has coordinated a high level policy group for the purpose of improving foreign compliance with Australian management measures.

38 This Table has been compiled from Governmental statements and Ministerial Press Releases, principally from the Australian Department of Fisheries, Forestry and Agriculture, the Australian Fisheries Management Authority and the Customs Department.

Chapter 8
Australia's Response to IUU Fishing

8.1 Introduction

This Chapter addresses the ways in which the Commonwealth Government has responded to the problems raised by persistent IUU fishing within the Heard and McDonald Islands' Fishing Zone. The measures adopted by Australia have been categorised to reflect the range of measures identified within the *IPOA-IUU* for adoption by coastal States and national governments. These measures were discussed in detail in Chapter 4. Particularly since 1999, when the force of the problem created by IUU fishing became more evident, the Australian response has been to adopt a coordinated legislative and policy approach to the problem. This is appropriate for it has often been acknowledged that there is no one measure that can be employed to successfully deter IUU fishing.[1]

8.2 Australia's Surveillance and Enforcement Policy

The *IPOA-IUU* states that through 'comprehensive monitoring, control and surveillance of fishing from its commencement, through the point of landing, to final destination…'[2] all States can act to deter IUU fishing. The Australian Commonwealth fisheries policy has been continually revised and refined in response to both the persistent nature of IUU fishing and the increasingly organised character of the IUU operator and fisher. It involves a combination of monitoring and surveillance through the implementation of the Southern Ocean Policy as well as control achieved through legislative provisions.

8.2.1 Responding to an escalating threat

The Australian Southern Ocean Surveillance Program was established in 1998 and is implemented by the Australian Fisheries Management Authority.[3] The

---

1 This was acknowledged during discussions at the *OECD Workshop on IUU Fishing Activities*, Paris, 19-20 April 2004. <www.oecd.org/documents/5/0,2340,en_2649_33901_21007109_1_1_1_1,00.> (May 2004)

2 *IPOA-IUU*, Section IV, paragraph 24.

3 The phrase 'Monitoring, Control and Surveillance' is often employed when discussing initiatives or policy involved in the surveillance of fisheries.

Program relies upon the use of civilian patrol vessels, and the cost of conducting fisheries enforcement in the remote Heard and McDonald Islands' AFZ is significant. The Australian Government committed AU$15.8 million over four years from the 1998/1999 fishing year to June 2003.[4] In May 2003, an additional AU$10.8 million was allocated in the 2003/2004 budget to enhance the capability of patrols.[5] A single patrol from mainland Australia to the Heard and McDonald Islands' AFZ is estimated to cost AU$2 million.[6]

As the majority of arrests have involved an element of the Australian Defence Force, the budgeted expenditure for surveillance and patrol has often been exceeded. During the trial of two crew members of the *Salvora,* the first foreign fishing vessel arrested by Australia, the prosecutor informed the District Court of Western Australia that:

> Two large vessels HMAS *Anzac* and *Westralia* were required to travel a significant distance and handle some rough sea conditions to arrive at about 4000 kilometres south-west of Fremantle where the *Salvora* was boarded. The cost of running the two ships is significant and the HMAS *Anzac* carried a helicopter which was used in the boarding. In addition, two RAAF aircraft undertook forward surveillance at a significant cost.[7]

Notwithstanding a governmental direction that Australian Defence Force personnel and hardware were only to be used as a last resort if civilian enforcement agencies were unable to secure an arrest,[8] elements of the Defence Force have been employed in seven of the eight arrests to date.[9] This position is reflected in the 1997 Australian Strategic Defence Policy which states that:

> Offshore territories require no special consideration in Australian defence policy, except for the need to be able to assist other

---

4 Australian Fisheries, Forestry and Agriculture Fact Sheet: 'Global Fisheries Issues affecting Australia', undated <www.affa.gov.au > (August 2001) (copy on file).

5 I. Macdonald, Minister for Fisheries, Forestry and Conservation and C. Ellison, Minister for Justice and Customs: ' $12 million budget boost to fight illegal fishing in Southern Ocean' 13 May 2003. $10.8 million is directed to surveillance and $1.8 million to improve post arrest processes.

6 S. Morris and A. Hodge, "Cold pursuit finally reels in Toothfish poachers' *The Australian* 29 August 2003, 3. The high costs involved are also confirmed in the judgment of *The Queen v Santome* and *Paz* (Unreported, Western Australian District Court, O'Sullivan J, 14 October 1998), 16.

7 *The Queen v Santome* and *Paz* above, 16.

8 Australian Fisheries Management Authority Fact Sheet: *The AFMA Southern Ocean Surveillance Program,* undated, <www.afma.gov > (copy on file).

9 See Table 8.4. The *Viarsa* was arrested following a 21 day pursuit by Australian Customs Officers supported by British and South African vessels. The other arrests all involved the use of the Australian Defence Force in some way.

10 Australia's Strategic Policy 1997, 30

agencies from time to time with surveillance and enforcement operations in the EEZs surrounding those territories.[10]

The Surveillance Program sets a monitoring schedule in advance which can be altered to meet changing circumstances. The aims of the Surveillance Program are to:

- deter illegal fishing operations by maintaining a presence in the Heard and McDonald Island's Australian Fishing Zone;
- warn off possible illegal fishermen/vessels operating on the high seas adjacent to the Heard and McDonald Islands' Australian Fishing Zone;
- initiate hot pursuit with recourse to the Australian Defence Force for assistance where needed;
- collect illegal fishing gear if found inside the Australian Fishing Zone; and
- gather intelligence on possible offenders.[11]

On 25 July 2002, the Minister for Fisheries, Senator Ian Macdonald, announced plans to: '[G]et tougher on Toothfish poachers'.[12] Exact details of the enhanced program were not available for public release, but the Government announced that patrols would be increased within the Heard and McDonald Islands' Fishing Zone.[13] Other measures outlined included cooperating (through possible joint surveillance agreements) with relevant nations including France, South Africa, the United Kingdom and New Zealand; developing cooperative arrangements with relevant trading countries and working with these countries to restrict market access for illegal catches of Toothfish,[14] and using strong diplomatic efforts to exert pressure for better controls over illegal fishing.[15]

---

11 Australian Fisheries Management Authority, Fact Sheet, *Operation Cosmo - Hot Pursuit and apprehension of suspected Illegal Fishing Vessel* undated.

12 I. Macdonald, Minister for Fisheries, Forestry and Conservation, 'Government to get tougher on Toothfish poachers' 25 July 2002. <www.mffc.gov.au/releases/> (15 August 2002).

13 The frequency of patrols and details as to when, where and who they will be conducted by, are not available to the public. For obvious security reasons, the Minister's Office would not comment on whether Australian Defence Force members would form part of the patrols.

14 This measure is linked to the move by Australia to support the listing of the Toothfish under CITES at the Meeting of *CCAMLR* Convention Parties in November 2002. Successful listing would impose strict trade regulations on the import of Toothfish. Verification that the trade is legal would be required. Although the proposed listing was not supported by *CCAMLR* parties, Australia views a CITES listing as complimentary to the *CCAMLR* management measures. See S. Stone, Parliamentary Secretary to the Minister for the Environment and Heritage, 'Major Blow to Illegal Fishing' 11 June 2002, < www.ea.gov.au/minister/ps/2002/psmr11june 202> (November 2002) S. Stone, 'Australia continues to derive Protection for Toothfish' 2 November 2002, <www.ea.gov.au/minister/ps/2002/psmr2November 202> (November 2002).

15 Macdonald, 'Government to get tougher on Toothfish poachers', above.

The resolve of July 2002 was followed by a February 2003 announcement of a Federal Cabinet agreement to an 'intensive campaign to protect our fisheries around Heard Island and McDonald Island'.[16] To this end, in May 2003 the Commonwealth Government announced that the civilian charter vessel program would be replaced by armed enforcement patrols. The vessels were to be manned by armed Customs officers providing support to Australian Fisheries Management authority officers with no immediate plans to introduce an Australian Defence Force element into patrols.[17]

The first armed civilian fisheries patrol in the Heard and McDonald Islands' AFZ was conducted between April and May 2003. Australian Customs officers were armed with 9mm Glock pistols, handcuffs, capsicum spray and batons. M16 rifles were also available if required.[18] In an environment where foreign fishing vessels have repeatedly displayed an intention to run rather than submit to arrest,[19] this armoury seems lightweight. Pistols, handcuffs and sprays are effective at close quarters, once the fleeing vessel has actually been forced to stop and been safely boarded. The patrol leader was interviewed on the vessel's return to the Australian mainland and was quoted as stating that, '[I]f a foreign fishing vessel had been intercepted, suitable weather conditions would have been needed to board'.[20]

In the Southern Ocean, 'suitable' weather conditions are rare and the chances of a successful boarding slight. Indeed the *Maya V* was apprehended in January 2004 with the assistance of a Naval boarding party from the HMAS *Warramunga*. The sailors were 'fast-roped' on board the *Maya V* from the ship's helicopter.[21]

8.2.2 More 'force' in enforcement or alternative measures?

In the course of coastal States enforcing resource management laws, it is not appropriate that person's lives are at risk or lost. This is a view the Australian Department of Defence concurs with.[22] Indeed the Australian Government has responded to the increasingly bold nature of IUU fishers (as outlined in Part 2) through a number of policy and legislative measures as examined in this Chapter, none of which involve placing the lives of Australian customs officers or Defence

16 I. Macdonald, Minister for Fisheries, Forestry and Conservation, 'Australia steps up the battle on illegal fishing' 9 February 2003 <www.mffc.gov.au/releases> (14 February 2003).

17 Macdonald and Ellison, '$12 million Budget boost to fight illegal fishing in the Southern Ocean', above.

18 Waterhouse, 'No bites on armed fisheries patrol' *The Mercury, 13* May 2003.

19 As supported by the three recent hot pursuits of vessels detected within the Heard and McDonald Islands' AFZ .

20 Waterhouse, 'No bites on armed fisheries patrol', above.

21 I. Macdonald, Minister for Fisheries, Forestry and Conservation and R. Hill, Minister for Defence, 'Navy catches suspected Illegal Fishing Vessel' 24 January 2004, <www.affa.gov.au/ministers/macdonald/releases/2004/04015mj.html> (6 May 2004).

personnel at risk nor, for that matter, the lives of the IUU fishers. However, the fact remains, that the act of apprehension of IUU fishing vessels on the Southern Ocean is an inherently dangerous act with risk involved.

Given the recent commitment in the Federal budget of May 2005 to armed patrols until 2010, it is recommended that the viability of the armed patrols be examined by the Government with the aim of identify means of maximising the enforcement capability of the civilian armed patrols yet also acknowledging that it would be undesirable at both the national and international level for more force to be invoked against IUU fishers.

In these circumstances, it is suggested that there are more efficient ways to combat IUU fishing. Pursuing IUU fishing vessels through a vast and hostile Southern Ocean is a costly and ineffective use of civilian and defence resources. This is particularly so when the lightweight armoury seems unsuitable for the marine environment in which it is deployed. There are alternative (and more cost effective) measures to deter IUU fishing which don't involve the use of force and which can be applied at the point of landing or the sale of the illegally begotten catch. Such measures applied through port and market States have been adopted by *CCAMLR* and are applied by Australia as a *CCAMLR* contracting party.

This is not to suggest that monitoring and surveillance should be abandoned. The patrols by the *Southern Supporter* and now the *Oceanic Viking* have been instrumental in deterring the presence of the IUU fleet which in the late 1990's reportedly numbered up to three score. What is needed are complementary measures to assist the surveillance program. In the event that a vessel is detected fishing without authorisation within the Heard and McDonald Islands' AFZ, but not apprehended, its presence can be communicated to *CCAMLR* members and cooperating non-contracting parties with the aim of stopping the vessel from unloading its illegal catch at a member port or from successfully trading that catch.

8.2.3 The 2004/2005 Southern Ocean Policy

As part of the May 2005 Federal Budget, the Commonwealth Government announced that 'regular armed patrols of remote Australian waters in the Southern Ocean will continue until at least 2010.[23] To support the patrols, funding of up to AU$217.2 million will be made available. This funding announcement followed closely upon the commitment of AU$89.2 million to fund armed patrols in 2004 and 2005.

---

22 Australian Fisheries Management Authority, *The AFMA Southern Ocean Surveillance Program*, above and Australia's Strategic Policy, above.

23 I. Macdonald, Minister for Fisheries, Forestry and Conservation, 'Long-term Commitment to Southern Ocean armed patrols' 10 May 2005, <www.mffc.gov.au/relesaes/2005/05078mj.html > (20 May 2005).

by a subregional or regional fisheries management organisation that Australia is a member of or participates in.[36]

Most importantly, the *Fisheries Management Act* was amended in 1999 to authorise Australian Fisheries Management Authority officers to take action against vessels flagged to member States of the *UN Fish Stocks Agreement* on occasions where those vessels were fishing on the high seas but in an area covered by a regional organisation or arrangement involving Australia.[37]

In 2004 the ratification of the *Compliance Agreement* necessitated further amendment to the two principal Acts regulating Commonwealth fisheries. Schedule 1 of the *Fisheries Legislation Amendment (High Seas Fishing Activities and Other Matters) Act 2004* (Cth), implements the *Compliance Agreement* into domestic law. Sections 6 and 7 of the *Fisheries Administration Act* (which were amended with the implementation of the *UN Fish Stocks Agreement*) were further amended.[38] Section 6 refers to Australia's international obligations in the terms below.

Fisheries Administration Act 1991 (Cth), Section 6

(ba) ensuring that:

(i) the exploitation in the Australian Fishing Zone (as defined in the *Fisheries Management Act* 1991) and the high seas of fish stocks in relation to which Australia has obligations under international agreements; and

(ii) related activities;

are carried on consistently with those obligations; and

(bb) to the extent that Australia has obligations:

(i) under international law; or

(ii) under the *Compliance Agreement* or any other international agreement;

in relation to fishing activities by Australian flagged boats on the high seas that are additional to the obligations referred to in paragraph (ba) – ensuring those activities are carried on consistently with those first mentioned obligations.

---

36 *Fisheries Legislation Amendment Act* (No. 1) *1999* , Section 4 amended Section 7 of the *Fisheries Administration Act 1991* (Cth).

37 *Fisheries Legislation Amendment Act* (No. 1) *1999* (Cth) Section35. This incorporates the provisions of Article 21 of the *UN Fish Stocks Agreement* which provides for third State boarding rights as

The *Compliance Agreement* also calls upon flag States not to grant fishing authorisations to any fishing vessel which has been previously registered in the territory of another State and has previously undermined conservation and management measures.[39] The Australian government has given effect to this duty through the introduction of a new restriction on the granting of fishing concessions by the Australian Fisheries Management Authority.[40] Fishing authorisations are not to be granted to Australian flagged-vessels to fish on the high seas in circumstances where the vessel has previously been registered and authorised to fish in a 'foreign country' and the Australian Fisheries Management Authority has reasonable grounds to believe that the owner or operator of the vessel has undermined international conservation and management measures and the foreign country has suspended or cancelled the authority to fish.[41]

A further significant amendment to the Commonwealth fisheries legislation creates a requirement to establish and maintain a High Seas Register 'containing particulars of all Australian flagged vessels that are authorised from time to time to engage in fishing activities on the high seas.'[42] The details required to be included in the High Seas Register reflect the requirements of the *Compliance Agreement*.[43] The Australian Fisheries Management Authority is to provide the FAO with a copy of the contents of the High Seas Register[44] and this further supports the obligation in the *Compliance Agreement* on the FAO to establish a High Seas Vessel Authorisation Register (HSVAR). Finally the Australian High Seas Register is to be publicly available (on payment of the prescribed fee).[45]

against vessels flagged to member States to the *UN Fish Stocks Agreement* notwithstanding that the flag States is not a member of the relevant regional organisation or arrangement.

38 As was section 3 of the *Fisheries Management Act* which mirrors section 6 of the *Fisheries Administration Act*.

39 *Compliance Agreement*, Article III(5).

40 *Fisheries Legislation Amendment (High Seas Fishing Activities and Other Matters) Act 2004* (Cth) new Section 16B introduced.

41 The Section further stipulates, in relation to the suspension or cancellation, that the period of suspension has not expired or a period of three years has not elapsed since cancellation. Under section 16B (4) the Australian Fisheries Management authority may consider evidence that the owner or operator of the fishing vessel, at the time of suspension or cancellation has no beneficial or financial interest in, or control of, the vessel at the relevant time.

42 *Fisheries Legislation Amendment (High Seas Fishing Activities and Other Matters) Act 2004* (Cth), new Part 4A- Register of vessels authorised to fish on high seas.

43 *Compliance Agreement*, Articles VI (1) and (2). The *IPOA-IUU*, paragraph 42 provides further elaboration. See Table 4.1 in Chapter 4 for a list of particulars required. Section 57B of the *Fisheries Management Act 1991* (Cth) lists the particulars required for the new High Seas Register.

44 *Fisheries Management Act*. Section 57C.

45 *Fisheries Management Act 1991* (Cth). Section 57D.

8.4 Review of Fisheries Legislation

8.4.1 Introduction

Commonwealth legislation applicable to the management of fisheries has been amended on several occasions since 1999 to respond to the growing threat IUU fishing poses to Australian interests, not just within the Heard and McDonald Island's fishing zone, but within the wider Australian EEZ. The *IPAO-IUU* calls for all States to amend their national legislation[46] and in order to address the drivers of IUU fishing (as identified in Part 2) it is necessary that any legislative review increase penalties and lessen the economic benefits of being involved in IUU fishing.

Australian courts have consistently acknowledged that 'the proper management of Australian marine resources is undermined by IUU fishing offences and that there is a special need to deter offenders where surveillance and detection are difficult and costly'[47]

As discussed below, penalties for various fisheries offences applicable to foreign fishing vessels have been reviewed and increased in 1999 and again in 2004.[48] Strict liability offences and forfeiture provisions for arrested foreign fishing vessels were also introduced in 1999. Two new statutory provisions, specifically targeting the persisting problem of illegal fishing, were implemented in 2003. First, the Commonwealth may now recover, as a debt, the costs of any hot pursuit undertaken to secure the apprehension of a foreign fishing vessel. Whilst the intention of this amendment is commendable, the practicalities of actually recovering any monies from the owners of arrested foreign fishing vessels may prove frustrating and incur yet more costs.[49] The second amendment alludes to the difficulties encountered in policing the remote and hostile Southern Ocean and relates to the requirement that Commonwealth officers show identification to foreign vessels.

The effectiveness of the above mentioned amendments has, with the exception of the forfeiture provisions, not been tested. The main obstacle to the legislative impact in terms of enhancing Australia's enforcement capabilities within the Heard and McDonald Islands' Fishing Zone (and in deterring IUU fishing), lies in the very

---

46 *IPOA-IUU*, paragraphs 16-17 and 21.

47 M. White and S. Knight, 'Illegal fishing in Australian Waters- the use of UNCLOS by Australian courts' 11 (2005) *Journal of International Marine Law* 110, 118 and n 80. The authors refer to the case of *Mackay v Diman* [1999] NTSC 96 (unreported 7 September 1999, Riley J) which lists relevant authorities supporting the statement.

48 The fact that illegal fishing is conducted by Australian vessels within the Australian EEZ is not discounted, however the focus of this work is the implementation of legal measures to deter IUU fishing within the Southern Ocean. There is not currently a problem with Australian owned and flagged fishing vessels operating illegally in the Heard and McDonald island fishery, Australia's only Southern Ocean fishery, and thus a review of laws applicable to Australian vessels is not relevant.

49 This aspect is examined further below in Section 8.5.1.

nature of illegal fishing. The industry is controlled by highly organised corporate entities driven by profit margins.[50] The Commonwealth must expect the *Fisheries Management Act* to be challenged at every opportunity.[51] Whether it can withstand these challenges is a matter to be tested on the individual facts, and ultimately if the validity of the legislation is challenged, this remains a matter for the Courts. That said, the most recent amendments to the Act evince a clear commitment on the part of the Commonwealth to construct a rigorous regulatory framework which is not limited to simply increasing applicable maximum fines.

8.4.2 Fisheries offences applying to foreign fishing vessels

Division 5, Part 6 of the *Fisheries Management Act* outlines the offences pertaining to foreign vessels. A 'foreign vessel' is defined as a 'vessel other than an Australian vessel.'[52] The central element in many of the offences is: 'fishing within the AFZ without a foreign fishing licence.' Section 34 of the *Fisheries Management Act* governs the granting of such licences. Fishing has been broadly defined in section 4 of the Act to mean:

(a) searching for, or taking fish; or

(b) attempting to search for, or take fish; or

(c) engaging in any other activities that can reasonably be expected to result in the locating, or taking, of fish; or

(d) placing, searching for or recovering fish aggregating devices or associated electronic equipment such as radio beacons; or

(e) any operations at sea directly in support of, or in preparation for, any activity described in this definition; or

(f) aircraft use relating to any activity described in this definition except flights in emergencies involving the health or safety of crew members or the safety of a vessel; or the processing, carrying or transhipping of fish that have been taken.

---

50 This statement is based upon the evidence that IUU operators are increasingly organised in the conduct of their activites. See Part 2, Chapter 3.

51 As it has been in the past. The Master of the *Big Star* appealed the fine of $100,000 imposed when convicted on charges under ss100(1) and 100(A) of the *Fisheries Management Act.* His fine was reduced to $24,000. See *Perez v The Queen* [1999] 21 WAR 477. The Master of the *South Tomi* pleaded not guilty to a charge under s108(c) – refusing or neglecting to comply with an order given under s84 and was acquitted. Evidence was led by the defendant that the order was non specific, requiring the vessel to 'head to port' and as such could not be complied with Per personal communication with AFMA Officers 14 February 2002. Most recently the owners of the *Volga* appealed the decision of the Federal Court which upheld the validity of the forfeiture provisions in the *Fisheries Management Act*. See *Olbers v. Commonwealth (No.4)* (2004) 205 ALR 432. (*Olbers No.4*)

52 *Fisheries Management Ac 1991* (Cth), Section 4. The term 'vessel' is used in the *LOSC*. See, eg Article 292. The International Tribunal for the Law of the Sea (ITLOS) also adopts the term 'vessel'. The *LOSC* does use the term 'ship' in some Articles. See eg, Article 17 on the right of innocent passage.

Fisheries Management Act 1991 (Cth), Section 106,
(before the 1999 amendment)

Upon a conviction of a person under sections 95, 99 or 100 the court may order the forfeiture of all or any of the following:

(a) the vessel, net, trap or equipment used in the commission of the offence;

(b) fish on board such a vessel at the time of the offence;

(c) the proceeds of the sale of any such fish.

The Commonwealth's interest (after the arrest but before any conviction of crew) in the *Aliza Glacial* in 1998 was, therefore, no more than a potential interest, and consequently, subject to the existing property rights of a mortgagee. Both of the crew members charged with offences - Captain Andreassen and Fishing Master Miranda - had left Australia and there was little likelihood of either defendant returning to face the charges. This fact was relevant to Justice Ryan's observation that he was not inclined to delay the order sought by the mortgagee for the sale of the vessel.

The need for an actual order of forfeiture is reflected in the legal process followed in relation to the arrest of the *Big Star* in 1998. Master Perez was convicted under sections 100 and 100A of the *Fisheries Management Act* and a court order was made for the forfeiture of the vessel to the Commonwealth. However, as the vessel had already been released on a bond settled under Article 73 of the *LOSC*, the vessel was not able to be recovered. To date, the Commonwealth has been unable to exercise its proprietary rights.[68]

The intention of the Government that their legitimate interests in arrested foreign fishing vessels not be defeated again is apparent in the Second Reading Speech of the *Fisheries Legislation Amendment Act (No1),* which introduced sections 106A-106H. The Minister for Fisheries stated:

68 *Perez v The Queen*, [1999] 21 WAR 477. Per Owen J who noted the security documents were executed on 14 May 1998 and the vessel sailed from Fremantle that day.

69 Senate Hansard, Second Reading Speech for the Fisheries Legislation Amendment Bill 1999. Minister for Agriculture, Fisheries and Forestry. House of Representative Hansard, 1 September 1999, 9565.

> The amendment makes clear third party interests will not prevail over Commonwealth enforcement action by virtue of the *Admiralty Act.* Amendments under schedule 1 will provide for a more effective catch, gear and vessel forfeiture scheme to deter illegal fishing in the Australian fishing zone.[69]

In order to remove any possible doubt over the priority of the Commonwealth's interest in vessels seized under the new section 106A, and to perhaps avoid a repeat of the political embarrassment of 'losing' the *Aliza Glacial* to a Norwegian bank,[70] section 108A was also inserted by the 1999 amendments. This section is reproduced below.

Fisheries Management Act 1991 (Cth), Section 108A

(1) The seizure, detention or forfeiture of a vessel under this Act has effect despite any or all of the following events:

(a) the arrest of the vessel under the Admiralty Act 1988;

(b) the making of an order for the sale of the vessel by a court in proceedings brought under the Admiralty Act 1988;

(c) the sale of the vessel under an order made by a court in proceedings brought under the Admiralty Act 1988.

(2) Subsection (1) has effect regardless of whether the seizure, detention or forfeiture, or the event that was the basis for the seizure, detention or forfeiture, occurred before or after the arrest, making of the order or sale (as appropriate).

To appreciate why the Government has been so intent on making the enforcement and forfeiture provision in the *Fisheries Management Act* as water tight as possible, one needs to reflect on the problem of illegal fishing. Although section 106 had been in force since 1991, it was not tested until the *Aliza Glacial* litigation in 1998. When the legislative provisions were found wanting, the Government took decisive steps to ensure that future illegal foreign fishing vessels could be validly seized and

70 Senate Hansard, 8 July 1998, 5229. Senator Murphy was critical of the government for 'allowing a Norwegian bank to repossess the vessel' and noted the government was 'liable for the costs associated with this fiasco.' At the time the Senator inquired as to the 'steps the government will be taking to ensure that in the future we are not confronted with this sort of situation.'

forfeited under the *Fisheries Management Act*. The legal rights of the mortgagee in the event that a vessel owner defaults on loan repayments seem to have been overlooked by the Commonwealth. Through the implementation of automatic forfeiture provisions in 1999, the mortgagee loses priority in any proprietary claim over the vessel and must seek to recover any monies due as an unsecured debt. This result defeats the objective underpinning the law of securities and whilst the reason for the legislative amendment is supported, the impact requires reconsideration.

These amended forfeiture provisions were challenged in 2002 when Olbers Ltd, the owners of the *Volga,* availed themselves of their rights under sections 106E and 106F of the *Fisheries Management Act* to claim the vessel and institute legal proceedings.

The *Volga* litigation

The *Volga* was apprehended on 7 February 2002 and was something of a bonus to authorities who were pursuing the *Lena*.[71] Olbers Co. Ltd commenced proceedings in the Federal Court on 21 May 2002 challenging the validity of the forfeiture provisions under sections 106A-106H of the *Fisheries Management Act.* The main thrust of their argument was that before section 106A could operate to effect a forfeiture of the vessel, the gear and catch on board; it was necessary that there be a conviction for one or more of the offences upon which such forfeiture was said to based. This line of argument is similar to the reasoning of Ryan J in the *Aliza Glacial* litigation, however in the former case the legislation supported such an argument. The post 1999 legislation does not require either a conviction for fisheries offences or a court order to make the forfeiture effective.[72]

The parties to the *Volga* case appeared before Justice French on three occasions prior to his determination of the substantive issues.[73] In *Olbers* (No 4) French J referred to the opportunity for the owners of forfeited vessels to contest the forfeiture under section 106F, and concluded:

> Absent the institution of such proceedings within thirty days of a notice of seizure under s 106C the asserted forfeiture will be put beyond question by operation of s 106E. That process requires no conviction to have been recorded. I reject the contention that s 106A depends for its application upon a conviction for one or more of the offences mentioned in it.[74]

---

[71] The *Lena* was arrested on 6 February 2002 after previously evading arrest in December 2001.

[72] This is the finding of French J in *Olbers Co Ltd v The Commonwealth of Australia* (No 4) 205 ALR 440 and the Full Federal Court in *Olbers Co Ltd v The Commonwealth of Australia* [2004] FCAFC 262 (unreported J of Black CJ, Emmett and Selway JJ)

[73] *Olbers Co Ltd v Commonwealth of Australia* [2002] FCA 1269, 16 October 2002. This matter involved a request for security for costs by the Commonwealth which was not granted. *Olbers Co Ltd* v *Commonwealth of Australian* (No 2) [2003] FCA 177, 11 March 2003 involved a request by Olbers for a stay of proceedings pending the disposition of criminal charges against crew members

The applicants appealed the decision of Justice French. In September 2004 the Full Federal Court upheld the lower court decision.[75] In essence, the appeal judges concurred with the view of French J, that the vessel was forfeited to the Commonwealth upon commission of the offence. Officers boarding the vessel were therefore acting as agents of the Commonwealth, the new owners of the vessel.[76] Special leave to appeal to the High Court of Australia was refused in April 2005.[77]

The win at first instance was heralded as a victory for the Commonwealth Government, in that the intended effect of the legislation was confirmed. Shortly after Justice French dismissed the application by Olbers Ltd, the Minister for Fisheries stated:

> ...in the epic legal process that Olbers have pursued, the Government has shown its determination to uphold Australian law to defeat pirate operations in our territorial waters around Heard Island and McDonald Islands ...This is now the third legal case that the owners of the Volga have brought against the Commonwealth ... On each occasion the courts have decided that the Australian Authorities have acted correctly. Yesterdays' landmark ... decision ...supports the Government's view that if a foreign vessel is sighted illegally fishing in Australian waters then that vessel, its equipment and catch is automatically forfeited to the Commonwealth and becomes the property of the Commonwealth.[78]

8.4.4 The effectiveness of the Forfeiture provisions

Upon forfeiture, the scuttling of the foreign fishing vessels is proving to be an effective method of removing them from the illegal fishing industry. The evidence has shown that released vessels are quickly re-equipped and sent back to the Southern Ocean fishing grounds by their corporate owners.[79] The case of the *Salvora* illustrates the point.

---

of the *Volga*. The stay was refused. *Olbers v Commonwealth of Australia* (No 3) [2003] FCA 651, 26 June 2003 involved a motion by Olbers for a separate trail on four issues of law. Justice French determined that it was not appropriate to have a separate trial, see paras 30-38.

74 Olbers (No 4) above, 440.

75 *Olbers Co Ltd* v *Commonwealth of Australia* [2004] FCAFC 262.

76 Ibid, paragraph 22.

77 Special Leave Application, *Olbers Co Ltd* v *Commonwealth of Australia* [2005] HCA Transcript 228 (HCA Hayne and Callinan JJ, 22 April 2005

78 I. Macdonald, Minister for Fisheries, Forestry and Conservation 'New Chapter in maritime law- Attempt to claim back the *Volga* rejected ' 13 March 2004, <www.mffc.gov.au/releases/2004> (15 March 2004).

79 ABC News Online, 'Court dismisses poachers' vessel appeal' 13 March 2004. <www.abc.net.au>.

The *Salvora* was arrested in October 1997 by Australian authorities. In the months prior to her arrest she reportedly unloaded three separate catches of Toothfish in Mauritius. Following the *Salvora's* release by the Australian authorities, the Vessel Monitoring System (installed on the vessel as a condition of the release) was switched off (or somehow became inoperative). The vessel was subsequently detected fishing illegally within South African waters off Prince Edward Island. The *Salvora* was arrested by the French for fishing illegally within French sub-Antarctic waters in May 2001.[80] Whilst the French authorities have a history of scuttling arrested foreign fishing vessels on the grounds of safety,[81] it is only in the last few years that Australian authorities have taken decisive action in relation to seized vessels. The fate of the eight vessels arrested to date is shown in Table 7.3 following.

Notices under section 106C of the seizure of the vessel, equipment and fish, have been made in relation to the arrest of the *South Tomi, Lena, Volga, Viarsa* and *Maya V*. The owners of the *South Tomi* instituted proceedings in the Federal Court to challenge the forfeiture order, however the application was withdrawn and the vessel, equipment and fish were condemned as forfeited.[82] The *South Tomi* was sunk off the Western Australian coast at Geraldton in late 2004 and will be used as an underwater diving attraction.[83] The *Lena* was sunk off Bunbury in 2003 to be used for similar purposes.[84] The owners of the *Maya V* did not challenge the forfeiture order within the statutory time frame and the vessel has been condemned as forfeited. It will be used for simulated boarding training in Western Australia.[85]

The confirmation of the validity of the forfeiture regime in the *Fisheries Management Act* by the Full Federal Court does raise several issues under international law. Whilst coastal States may 'in the exercise of sovereign rights to explore, exploit, conserve and manage the living resources in the EEZ, take such measures, including boarding, inspection, arrest and judicial proceedings'[86] it may not do so without observing applicable rules of international law. The coastal State is under a duty to promptly notify the flag State through appropriate channels in cases of arrest or detention.[87]

---

80 See, eg ISOFISH Fact sheet: '*Salvora*' undated and Greenpeace 'The Case of the *Salvora*' undated.

81 French authorities have in the past scuttled vessels on the basis that they represent either an environmental or safety hazard.

82 Per email conversation with AFMA Officer, Mr John Davis 14 February 2002.

83 'Fishing Future' Newsletter of the Australian Fisheries Management Authority, Vol 2, Issue 4, December 2004, 16.

84 Ibid, 13.

85 C.Ellison, Minister for Justice and Customs and I.Macdonald, Minister for Fisheries , Forestry and Conservation, 'New role for Toothfish Pirate' 21 May 2004. <www.mffc.gov.au> (25 May 2004).

86 *LOSC*, Article 73(1).

87 *LOSC*, Article 73(4).

The Federal court has found that the force of Australian law is to transfer ownership of foreign fishing vessels upon commission (or alleged commission) of a prescribed offence with no requirement for an actual conviction upon the presentation of evidence. It must follow from this reasoning that at the time of detection within the AFZ, a foreign fishing vessel becomes Australian property. If this is the case might it be argued that Australia is the new flag State? This line of reasoning raises many points of conflict between the application of national legislation and Australian duties under international law. These include the rights of the flag State to be notified of the arrest (assuming there is one). Furthermore, a system under which ownership can vest in coastal States on detection of an IUU vessel is open to abuse. In practice, Australia has complied with international obligations under the *LOSC* and has posted bonds and notified the flag State in accordance with Article 73.

An additional legal issue raised by the confirmation of the forfeiture provisions relates to the duty under Article 73(2) of the *LOSC*, to 'promptly release arrested vessels and crew upon the posting of a reasonable bond or security.' Even though there may be a forfeiture order under the *Fisheries Management Act*, it would seem that Australia's obligations under international law to release the vessel, on the payment of a reasonable bond, would prevail. Government authorities could not retain the bond and also exercise forfeiture rights under the *Fisheries Management Act*. It would also appear that even if the preference of the authorities was to retain possession of the arrested vessel and exercise proprietary rights vested by the operation of section 106A of the *Fisheries Management Act*, the payment of a bond under Article 73(2) of the *LOSC*, would preclude this. However if the owners of arrested vessels chose not to pay the bond set, then presumably the forfeiture could proceed. In instances where a vessel has already been forfeited and sunk, then the owners would need to institute proceedings under Part VI of the *LOSC* seeking, *inter alia*, compensation for the vessel, gear and catch on board.[88]

In late 2002, Russia, the flag State of the *Volga*, lodged an application with the ITLOS under Part XV of the *LOSC* for the prompt release of the *Volga*.[89] This was in addition to the Federal Court action also initiated in 2002 and reviewed above. It is sufficient to note the bond set by Australia under Article 73 of *LOSC* was reduced by ITLOS. However, the bond has not yet been paid by the vessel's owners and the *Volga* remains, at the time of writing, tied up in Fremantle.[90]

88 This is assuming the bond has been paid under article 73. An application for prompt release in this instance would not be appropriate.

89 The *Russian Federation v Australia* (the *Volga* case) Case No.11 ITLOS, Judgement dated 23 December 2002.

90 ITLOS set the bond at AU$1,920,000 down from the original amount calculated by Australia of AU$3,332,500. Russia has sought a bond of just AU$500,000. Ibid, paragraphs 53-54 and 95.

Vessel Name and date of arrest	Fate of vessel after apprehension
Salvora -16 October 1997	Released under bond with Vessel Monitoring System condition attached, Continued to fish illegally
Aliza Glacial -17 October 1997	Released to Mortgagee under Federal Court order
Big Star – 21 February 1998	Court order for forfeiture under section 106 (*before* 1999 amendments) following conviction of crew member. Vessel had already been released on payment of bond with Vessel Monitoring System condition attached and failed to return.
South Tomi - 12 April 2001	Forfeited under s.106A *Fisheries Management Act* and sunk.
Lena - 6 February 2002	Forfeited under s.106A *Fisheries Management Act* and sunk
Volga -7 February 2002	Appeal to Full Federal Court challenging the validity and effect of sections 106A-106H of the *Fisheries Management Act* dismissed in September 2004. Special Leave to appeal to the High Court dismissed.
Viarsa -28 August 2003	Owners have filed application to challenge the notice of forfeiture under section 106C.
Maya V -22 January 2004	Forfeited under s.106A *Fisheries Management Act* and sunk

Table 8.2 Fate of foreign vessels apprehended in the Heard and McDonald Islands' Australian Fishing Zone

8.5 Further legislative responses

The amendments to the *Fisheries Management Act* passed in March 2004 via the *Fisheries Legislation Amendment (Compliance and Deterrence Measures and other Matters) Act 2004* and the *Fisheries Legislation Amendment (High Seas Fishing Activities and Other Matters) Act 2004*, are principally in direct response to the increasingly bold behaviour of illegal fishers. In addition to increased maximum fines, a number of changes have been introduced which are aimed at improving the overall effectiveness of the regulatory framework.

[91] Section 27 of the *Fisheries Legislation Amendment (Compliance and Deterrence and other Matters) Act* which inserts Subdivision CA into Division 6 of Part 6 of the principal Act. Section 106J-106S detail the procedure to be followed for the recovery of the costs.

8.5.1 The recovery of costs incurred in pursuit

One of the more significant amendments is the introduction of provisions for the recovery by the Commonwealth of the costs involved in the hot pursuit and apprehension of foreign fishing vessels.[91] The government decision to seek to recover the cost of pursuit and apprehension came shortly after the costly 21 day hot pursuit of the *Viarsa* in August 2003. The vessel was ultimately arrested with the assistance of both South African and British vessels. Australia is, reportedly, expecting to meet the costs incurred by South Africa and the United Kingdom in coming to her assistance.[92] The exact costs incurred are uncertain. The Minister for Fisheries has stated: '[T]he chase of the *Viarsa* was very expensive. The final figures are not in yet.'[93] The publicly available estimates range from, a statement that the cost is in the vicinity of four five, six, seven, eight or nine million dollars,[94] to the more definite statement made in the Second Reading Speech for the Fisheries Legislation Amendment (Compliance and Deterrence Measures and Other matters) Bill 2003, in reference to the arrest of the *Viarsa:*

> [A]fter the passage of this legislation, [costs will] be able to be recovered from that owners of the vessel...So the $4 million or $5 million costs that the Australian taxpayer was put to, to eventually apprehend that vessel will be able to be recovered in the future.

The procedural requirements for recovering the pursuit costs in relation to foreign vessels are contained in sections 106-106S with further details for working out the actual costs incurred by, or on behalf, of the Commonwealth to be prescribed by regulation. A reading of subsections 106L(1) and (2) shows three evidentiary issues to be satisfied before pursuit costs can be claimed as a debt. The first and third requirements are linked to the successful conclusion of the pursuit undertaken by authorities. First, the foreign vessel must be forfeited to the Commonwealth under section 106A. That is, it must have been used in one of the offences listed in that section. Second, the Master of the vessel must fail to either stop the vessel in accordance with orders under section 84(1)(aa) or bring the vessel to a place as directed under subsections 84(1)(k)or (l). Thirdly, if as a result of the failure, pursuit activities are undertaken with the result that the vessel arrives in Australia, the owner of that vessel is liable to pay to the Commonwealth by way of penalty, all pursuit costs incurred in respect of that vessel. [95]

The term 'pursuit costs' has been defined in section 106J by reference to costs reasonably incurred by or on behalf of the Commonwealth in respect of pursuit activities conducted in respect of a foreign vessel. This phrase is defined as meaning all costs:

---

92 Morris and Hodge, 'Customs closes in on poachers' *The Australian,* above and S. Morris and A. Hodge, 'Cold pursuit finally reels in Toothfish poachers' *The Australian,* 29 August 2003, 3.

93 Commonwealth of Australia, Senate Hansend, 10 February 2004, 19572

94 Ibid.

95 *Fisheries Management Act 1991* (Cth) Section 106L(1)(c).

(a) that the Commonwealth is liable to pay in respect of such activities; and

(b) that are directly attributable to the conduct of those activities;

and, without limiting the generality of the above includes:

(c) costs incurred by any Commonwealth agency or body in respect of such activities; and

(d) costs incurred by any arm of the Australian Defence Force that provides assistance in respect of such activities; and

(e) costs incurred by the government of any foreign country that provides assistance or facilities in respect of such activities, being costs so incurred on the basis that those costs will be reimbursed by the Commonwealth.

The intention that costs incurred by foreign governments assisting in the successful resolution of a hot pursuit is evident from the broad definition. The provision will also enable the Commonwealth to recover the costs of a pursuit which are additional to the routine patrols provided for in annual budget estimates. Thus, the figure of $AU1.23 million provided as an approximation of costs incurred over and above routine patrol costs in the apprehension of the *South Tomi* in 2001,[96] would be recoverable as a debt under the new legislation.

The recovery of pursuit costs works in a similar fashion to the forfeiture provisions examined above. A preliminary written notice of debt must be given to the Master of the vessel, or, if this cannot be conveniently done, the notice may be fixed to a prominent part of the vessel itself.[97] Full particulars of the pursuit costs are to be provided within 10 days of the preliminary notice of debt.[98] In the event that the owner fails to give notice of an intention to contest the debt claimed, within 30 days of receipt of the notice of full particulars, the debt becomes due and payable.[99] As with the forfeiture provisions in sections 106F-G of the *Fisheries Management Act*, the owner must institute Federal Court proceedings within two months for an order that the debt is not payable.[100] Section 106Q(1) specifies the two orders that may be sought - namely that the debt is not payable because the vessel was not forfeited to the Commonwealth or that the debt or part thereof was not reasonably incurred.

Finally, the burden of proof in establishing the evidentiary issues in section 106L and outlined above is split. The owner of the vessel is to establish on the balance of probabilities that the vessel is not forfeited to the Commonwealth, that is, that the vessel was not used in an offence against the sections listed in section 106A. The Commonwealth is to establish on the balance of probabilities that the Master of the

96 Commonwealth of Australia, Senate Hansard, Answer to Question on Notice, No. 730, 10 December 2002, 7,569.

97 *Fisheries Management Act (1991)* (Cth), Section 106M.

98 *Fisheries Management Act (1991)* (Cth), Section 106N.

99 *Fisheries Management Act (1991)* (Cth), Section 106P.

100 *Fisheries Management Act (1991)* (Cth), Section 106Q.

vessel failed to stop or bring the vessel to a place in Australia as directed and that successful pursuit activities commenced as a result.[101]

There have been no notices issued of recoverable debts yet for the last arrest in the Heard and McDonald Islands' AFZ was in January 2004. The are potential practical difficulties in realising any debt incurred in the pursuit of a vessel detected engaging in illegal fishing within the AFZ. First the vessel must be successfully apprehended and as the review above suggests, this usually requires a show of 'force'. Furthermore for the eight successful arrests over nine years, there are many many more vessels which have escaped arrest. Second, assuming a vessel is apprehended after failing to stop as ordered under section 84(1) of the *Fisheries Management Act*, posting a notice of debt and actually extracting the monies due from owners who usually shelter behind complex corporate structures are two distinct matters. It is unlikely the corporate owners of IUU vessels issued with such a notice will voluntarily hand funds to the Commonwealth government. If the vessel, gear and catch have already been forfeited to the Commonwealth through the operation of section 106A of the *Fisheries Management Act,* then there is no asset over which the debt can be secured.

8.5.2 Relaxation of obligation to show identification under section 84(6)

Prior to the amendment of section 84(6) of the *Fisheries Management Act*, a person required to do something under section 84(1) (such as stop a vessel or bring a vessel to a directed place) was only obliged to comply with the requirement if the officer produced written identification or an identity card. In the Southern Ocean, where the seas are wild and the weather extreme, producing such identification for inspection is problematic

Section 84(6) as amended and the newly inserted 84(6A) will allow for circumstances where it is impossible to produce written proof of identity. In such circumstances, the officer must produce identification at the first available opportunity, for example when boarding a foreign fishing vessel.

The rationale for this amendment appears to be linked to the amendments relating to the recovery of pursuit costs. As stated, the onus of proof to establish that the Master of the vessel did not comply with a section 84(1)(aa), (k) or (l) requirement in relation to the recovery of pursuit costs, is on the Commonwealth. If the owner of a foreign vessel can show that no identification was produced for inspection, notwithstanding the practical difficulties in doing so, there would be no obligation to comply with a section 84(l) requirement to stop the vessel. Correspondingly, no liability would arise to pay the costs incurred in any pursuit required to arrest the vessel.

---

[101] *Fisheries Management Act (1991)* (Cth), Section 106S.

In light of the demonstrated increased willingness of the owners of foreign fishing vessels to litigate in both international and domestic courts, the risk of a challenge based on non-compliance with section 84(6) is very real.[102] Following the amendment, the officer is able to produce the required identification on boarding a vessel without negating the legal effect of the section 84(l) requirement to stop. As a matter of caution the Australian government should remain vigilant to ensure this right is not abused and that the rights of vessels and flag States under international law are not infringed.

8.6 Port controls

In addition to supporting the conservation measures adopted by the *CCAMLR* Commission with respect to the CDS, Australia has participated at the international level to review port States measures to deter IUU fishing. Australian fisheries experts attended the *Expert Consultation to review Port States Measures to combat IUU Fishing*[103] and the *Technical Consultation to address substantive issues relating to the role of port States to prevent, deter and eliminate IUU Fishing.*[104] The *IPOA-IUU* specifically addresses port State measures in some detail

At the national level, foreign fishing vessels must obtain a port permit prior to entering an Australian port. The detailed information to be provided in the port permit application is prescribed by regulation.[105] Australia currently operates a closed port policy.

In January 2004 the Australian Government announced a review of catch landing arrangements for foreign fishing vessels. At the time it was also announced that until the review was completed 'new applications from foreign fishing vessels wanting to land or tranship their catch will only be granted in exceptional circumstances.'[106] Whilst at the time of writing the review had not been completed, the Australian Fisheries Minister has stated:

> A sensible and balanced port access and catch landing policy for foreign vessels could, for example, help us better monitor and control the movement of fish products that may have come from poaching activities.[107]

---

[102] See eg the sustained action by Olbers Ltd the owners of the *Volga* in the Federal Court and the application for prompt release to ITLOS by Russia, the flag State of the *Viarsa.*

[103] Held at the FOA in Rome in November 2002.

[104] Held at the FAO in Rome in September 2004.

[105] *Fisheries Management Regulations 1992* (Cth),Regulation 11.

[106] I. Macdonald, Minister for Fisheries, Forestry and Conservation, 'Review of catch landings in Australia' 29 January 2004. <www.mffc.gov.au/releases/2004> (29 March 2004).

[107] Ibid.

8.7 Market State Measures

The NPOA-IUU reports that currently, very small quantities of fish are landed in Australian ports from foreign fishing vessels and as such ,the 'actual market-related implications have to date been insignificant.'[108] Notwithstanding this fact, the Federal Minister responsible for Commonwealth fisheries announced a review of Australia's foreign catch landing arrangements in 2004.[109] At the time it was reported that the aim of the review would be to take a more restrictive approach to assessing new applications from people wanting to land or tranship catch from foreign fishing vessels.[110]

On a regional level, Australia is an active supporter of *CCAMLR* measures to limit opportunities for the trade in IUU fish. As discussed in Part 4, the Australian delegation proposed two primary amendments to the *CCAMLR* IUU vessel lists.[111] These amendments were implemented through an amended conservation measure[112] and should improve the effectiveness of the IUU vessel lists and correspondingly limit the opportunities for IUU fishing by raising presumptions of non-compliance for port and market States to act upon.

8.8 Publication of arrests

The *IPOA-IUU* calls upon all States to publicise widely, including through cooperation with other States, full details of IUU fishing and actions taken to eliminate it, in a manner consistent with applicable confidentiality requirements.[113] Each arrest of foreign fishing vessels within the Heard and McDonald Islands' fishing zone has been widely publicised by Australian authorities and hailed as a potent warning to other would be illegal fishers. In announcing the most recent arrest that of the *Maya V* in January 2004, the Minister for Fisheries, Forestry and Conservation stated:

> This joint Defence-AFMA apprehension marks another blow to illegal fishers and proves once again that Australia has the capacity to act decisively in all sorts of locations and conditions to protect our fisheries resources and territorial waters.[114]

---

108 AUS-NPOA-IUU, paragraph 95.

109 Macdonald, 'Review of foreign catch landings in Australia' above. No recent information on the status of this review is publicly available.

110 Ibid.

111 See Part 4.

112 Ibid.

113 *IPOA-IUU*, paragraph 32.

114 R.Hill, Minister for Defence, 'Navy catches Suspected Illegal Fishing Vessel' 24 January 2004.

8.9 Cooperation between States

8.9.1 Cooperative surveillance and enforcement

An alternative to high cost unilateral military surveillance and enforcement would be to commit to joint surveillance and enforcement with other coastal States under the framework of a regional treaty. The framework for bilateral or multilateral cooperation is already in place. The *UN Fish Stocks Agreement* specifically provides for the boarding and inspection of the flag vessels of member States, by authorised officers of other member States.[123] However, responsibility for the investigation of alleged offences and the imposition of penalties remains with flag States.

The *CCAMLR* System of Observation and Inspection also provides for the boarding and inspection of member's vessels within the *CCAMLR* Convention Area. This system is instituted on a national basis with inspectors appointed by contracting parties and is subject to the jurisdiction of the appointing State.[124] Reports submitted at the Commission's annual meetings however, indicate that number of inspections under the system has been very low in comparison to the number of designated Inspectors.[125] Ultimately inspections are reliant upon flag State consent to first, the inspection regime of the RFMO and second, to act in the event that further investigation of the fishing vessel is required.

Following the success of the cooperative efforts of the South African and Australian Defence Forces in apprehending the S*outh Tomi,* the then Australian Minister for Fisheries observed:

> It indicates the sort of national and international action required if we are going to prevent, deter and eliminate the scourge of illegal, unreported and unregulated fishing.[126]

---

[123] *The United Nations Agreement for the implementation of the provisions of the United Nations Convention on the Law of the Sea of 10 December 1982 relating to the Conservation and Management of Straddling Fish Stocks and Highly Migratory Fish Stocks,* opened for signature 4 December 1995, 34 ILM 1542 (entered into force 11 December 2001). Article 21.

[124] Text of *CCAMLR* System of Inspection, Paragraph 1(c) As adopted at *CCAMLR*-VII (1988) and amended at *CCAMLR*-XII (1993), *CCAMLR*-XV (1996), *CCAMLR*-XVIII (1999).

[125] D. Agnew, 'The Illegal and Unregulated Fishery for Toothfish in the Southern Ocean and the CCAMLR Catch Documentation Scheme' (2000) 24 *Marine Policy* 361, 364 and Table 2. See also, R. Rayfuse, 'Enforcement of High Seas Fisheries Agreements: Observation and Inspection under the Convention on the Conservation of Antarctic Marine Living Resources' (1998) 13(4) *International Journal of Marine and Coastal Law* 579, 590-591. Whilst it is not apparent why the number of inspections conducted has been low since the System's inception in 1988/90, the Commission has, on more than one occasion, called upon members to make greater use of the System. *CCAMLR*-XII (1993), paragraph 6.5.

[126] SAS hook fish pirates, *The Saturday Mercury,* 14 April 2001, 3

The benefits of international cooperation were once again made evident with the successful arrest of the *Viarsa* in August 2003. The Australian Government acknowledged:

> The assistance of these cooperating international vessels has been invaluable. Without them the operation would never have been as successful as it has been.[127]

8.9.2 Information Sharing

Whilst there have been instances of cooperative enforcement, the need for increased cooperative surveillance to detect (and deter) IUU fishing vessels is evident. For example, the *Viarsa* was active in the Southern Ocean before her arrest in August 2003. She reportedly unloaded 230 tonnes of Toothfish in Mauritius in June 2003. Monitoring of the movement of known IUU fishing vessels does occur at the industry level for this information was reported by COLTO.[128] A formal system at the government departmental level is required.

At the very least, a formal system of information exchange between States with sovereign interests in the Southern Ocean should be implemented. The evidence suggests IUU fishing vessels operating in the Southern Ocean do so in groups, alerting each other about the presence of Australian or French government patrol vessels in the area.[129] The Australian fishing industry has supported joint surveillance with other States as the only feasible alternative to deter the illegal fishing vessels.[130]

Strong informal working relationships have been forged at officer level between enforcement agencies in Australia, New Zealand, South Africa and France.[131] In March 2001, Australian Fisheries Management Officers visited South African Fishing Authorities to deliver presentations on Australian enforcement in the Heard and McDonald Islands' fishing zone and pursue opportunities for joint intelligence sharing.[132] In April 2001 the Australian authorities were able to call on their South African counterparts and request assistance in apprehending the *South Tomi* as it fled across the south Indian Ocean.[133]

---

127 Minister for Fisheries, Forestry and Conservation, *Bringing back the Viarsa* AFFA03/173MJ, 30 August 2003.

128 www.colto.org

129 Personal communication with Mr John Davis (AFMA Officer), 4 June 2001. See, also ABC, 'The Toothfish Pirates' 4 Corners, above..

130 Martin Excel, 'Exploration of Southern Ocean Fisheries: An Industry Perspective' in S. Bateman and D. R. Rothwell (eds) *Southern Ocean Fishing- Policy Challenges for Australia* (1998) Centre for Maritime Policy Paper No.7 103.111.

131 Personal communication with Mr John Davis, 4 June 2001.

132 Ibid . See also Report of Member's Activities in the Convention Area 2000-2001, Australian Report *CCAMLR* XX October 2001, Section VI(d)

133 Personal communication with Mr John Davis, 4 June 2001.

Whilst cooperation between the Southern Ocean coastal States remains informal (with the exception of the bilateral treaty with France) there are some avenues that could be pursued to establish a framework for systematic information sharing and cooperative surveillance. Shortly after the arrest of the *South Tomi* in 2001, the Australian Government stated:

> The successful outcome of this operation marked some significant firsts and has laid the groundwork for enhanced international cooperation in dealing with the problems of illegal fishing activity in both the remote waters of the Southern Ocean and other parts of the high seas. It is particularly significant to Australia given the distances involved and the environmental importance of the waters of the Southern Ocean.[134]

The international community has recently agreed to establish an Organisation for Economic Cooperation and Development Ministerial Taskforce to tackle the problem of illegal fishing.[135] Participating States include Australia, New Zealand, Namibia, Chile and the UK. Other identified opportunities for improved regional cooperation include are discussed below. In the broader context of establishing an international strategy for combating IUU fishermen, the Australian Government announced it would seek to enhance international agreements and increase diplomatic pressure.[136]

8.9.3 Bilateral treaty between Australia and France

Over several years, ongoing negotiations between Australia and France sought to settle the terms of a joint surveillance treaty to cover their respective maritime zones in the Southern Ocean. Both the Kerguelen Isles and the Heard and McDonald Islands are situated on the Kerguelen Plateau. On 24 November 2003, officials from both Governments signed the Treaty.[137] Under the terms of the Treaty the parties have agreed to exchange information on the location, movement and licensing of fishing vessels.[138] There is also provision for requesting the assistance of the other

---

[134] Statement by Minister for Fisheries, Forestry and Conservation, *Recent Government Responses to Illegal Fishing in the Southern Ocean* 7 June 2001.

[135] I. Macdonald, Minister for Fisheries, Forestry and Conservation, 'Pirates beware, Australia front and centre in the fight against illegal fishing' 2 December 2003. <www.mffc.gov.au/releases/2004> (2 December 2003).

[136] Macdonald, 'Government to get tougher on Toothfish poachers' above.

[137] I. Macdonald, Minister for Fisheries, Forestry and Conservation and A. Downer, Minister for Foreign Affairs and Trade, 'Maritime agreement sends a strong message to illegal fishing operators' (AFFA03/256MJ) 24 November 2003. See also, The *Daily Telegraph,* 'Canberra, Paris make up' 25 November 2003, 16. Treaty between the Government of Australia and the Government of the French Republic on cooperation in the maritime areas adjacent to the French Southern and Antarctic Territories (TAAF), Heard Island and the McDonald Islands. [2003] ATNIF 20.

[138] Ibid, Article 5.

party when engaged in a hot pursuit.[139] Whilst the Treaty is restricted to cooperative surveillance at this stage, it does contemplate the negotiation of further agreements on law enforcement operations.[140]

The May 2005 Federal budget provided funding for Fisheries officers to participate in French patrols in the Southern Ocean.[141] Fisheries Minister Macdonald has stated:

> Our cooperation with France will allow our Fisheries officers to frequently participate in French patrols and vice versa, in waters surrounding Australian and French Southern Ocean territories. This will effectively double the capacity of our present program in the Southern Ocean.[142]

8.9.4 Negotiations for cooperative surveillance with South Africa

Reference has been made to discussions on a draft text between Australian and the South Africa with a view to formalising cooperative action against illegal fishing within the Southern Ocean.[143] However, no formal arrangements have been implemented and both Governments rely upon the strong informal working relationships forged at the Departmental level.[144] As of August 2005 negotiations for a cooperative surveillance and enforcement treaty with South Africa for unspecified areas of the Southern Ocean were underway.[145] The ultimate aim of these negotiations is the formation of a bilateral treaty with South Africa, similar to the existing treaty with France, to further efforts to deter IUU fishing in the Southern Ocean.

8.9.5 Coalition of East Antarctic States

Members of the Australian Parliament have on occasion referred to a coalition of East Antarctic States (Australia, New Zealand, France, Norway and South Africa) however no formal arrangements have been settled.[146] The 1998 Australian Fisheries Management Authority report referred to a meeting of East Antarctic Coastal States in South Africa that year to address the 'considerable and increasing problem of illegal fishing for the Patagonian Toothfish in Antarctic waters.'[147] The specific aim of the

---

139 Ibid, Article 3(2) and (3).

140 Ibid, Annex III, Article 2.

141 Macdonald, 'Long-term Commitment to Southern Ocean armed patrols' above.

142 Ibid.

143 Senate Hansard, Answer to Question on Notice, 10 December 2002, above.

144 As evidenced in the arrests of the *South Tomi* and the *Viarsa*..

145 Email interview with Federal Attorney-General's Department, August 2005.

146 The 'Coalition' remains a loose association of States. Personal communication with Department of Foreign Affairs and Trade Legal Officers, 4 October 2001.

147 AFMA News, Vol 3 Issue 8, December 1999.

meeting was to discuss the types of measures that might collectively be taken.[148]

Although substantive outcomes were predicted from the 1998 meeting,[149] the Coalition remains informal and possibly unlikely to be pursued. In light of the bilateral treaty with France and the recent negotiations with South Africa it might be assumed that the Australian government's immediate preference is for bilateral treaties with neighbouring coastal States and that the bilateral treaty arrangements will facilitate an East Antarctic States cooperative regime.

8.10 Bilateral Hot pursuits

In recent years, Australia has conducted three hot pursuits of foreign vessels detected fishing illegally, all commencing within the Heard and McDonald Island's EEZ. Two ended successfully with the *South Tomi* being arrested in April 2001 and the *Viarsa 1* in August 2003. The *Lena* was arrested in February 2002 but not before an unsuccessful 14 day hot pursuit in December 2001.[150]

In addition to the budgeted funding, the recent hot pursuits conducted by Australian authorities have incurred considerable costs. It has been estimated that approximately AU$1.23 million was incurred over and above normal patrol costs in apprehending the *South Tomi* in 2001.[151] The *South Tomi* was detained after a 15 day pursuit which traversed 3,300 nautical miles across the Southern Indian Ocean. Australian Defence Force personnel boarded the vessel south of Cape Town with the assistance of the South African military.[152] The *Southern Supporter,* the civilian patrol vessel pursing the *South Tomi,* had been unable to force the vessel to stop. The government officers and civilian crew on board the *Southern Supporter* were unarmed and there was no helicopter boarding capacity.[153]

Whilst not confirmed, it has been reported that Australian expects to be billed for the assistance provided by South African and the United Kingdom during the pursuit of the *Viarsa.*[154] The *Viarsa* was pursued by the *Southern Supporter* for 21

---

148 Senate Estimates, Program 1.8 International Legal Interests – Foreign Affairs and Trade Committee 26 February 1998, 136.

149 Ibid.

150 These incidents were widely reported in the media. An archive of many of these media reports is maintained on the website of COLTO at: <www.colto.org/Articles.htm>.

151 Senate Hansard, Answer to Question on Notice, 10 December 2002, above.

152 Australian Fisheries Management Authority, 'Operation Cosmo' above. See, also, M. Fyfe, 'Poachers lose in the sail of the century' *The Age* , 6 May 2000, 3.

153 Ibid.

154 Morris and Hodge, 'Customs closes in on poachers' and 'Cold pursuit finally reels in Toothfish , *The Australian* above.

155 I.McDonald Minister for Fisheries, Forestry and Conservation, ' Dangerous Ice pursuit of *Viarsa* now longest in Australian maritime history', 25 August 2003 <www.mffc.gov.au/releases/2003> (23 August 2004).

days in August 2003, through treacherous and ice packed seas. [155] When ordered to accompany the *Southern Supporter* to port the *Viarsa* sailed west towards Cape Horn. Hampered by rough seas and poor weather, armed Customs officers on board the *Southern Supporter* were unable to effect an arrest.[156]

In late December 2001, the *Lena* evaded a civilian patrol vessel for 14 days before finally escaping with her alleged illegally caught hold of Toothfish.[157] She was later arrested in February 2002 with the aid of a helicopter boarding party from HMAS *Canberra*.[158] Table 8.4 below provides details of the three hot pursuits conducted by Australian authorities.

The right of hot pursuit is detailed in Article 111 of the *LOSC* and is one of the limited exceptions to the principle of flag State authority on the high seas.[159] The ability of the coastal State to exercise this right effectively is no easy task, however. Whilst the coastal State must take care to fulfil the cumulative procedural requirements of Article 111,[160] the vessels pursued are able to employ, and in practice have employed, a variety of practices which make pursuit difficult. In this context, the Director of Austral Fisheries recently made the comment that: "there needs to be force involved. It's not going to be, 'Welcome aboard and let me take you on a scenic cruise back to Fremantle.'"[161]

The *Viarsa 1* employed a range of techniques to avoid arrest during her pursuit in 2003. Tracking a dangerous trail through ice fields in an attempt to shake the pursuit vessel, *Southern Supporter*, the *Viarsa 1* was only deterred from following her track when advised that should she strike ice, by the time a rescue vessel arrived the crew would most likely have perished from hypothermia. Other evasive strategies adopted included shutting off communications, painting out the vessel's name (although the name was repainted during the 21 day pursuit) and tying ropes at regular intervals from high points to its railings for almost the entire length of the ship, presumably to deter helicopter boarding.[162]

---

156 Ibid. See also subsequent Media Releases AFFA03/152MJ dated 13 August, AFFA03/155MJ, dated 14 August, AFFA03/158MJ, dated 19 August.

157 Report of Member's Activities in the Convention Area 2001-2002 - Australia. *CCAMLR*-XXI (2002).

158 Ibid. In the weeks since the December 2001 hot pursuit, the *Lena* had been repainted and renamed the *Ana*.

159 These exceptions are listed in Art 110 and generally in under Part VII of the *LOSC* and include the right to board vessels engaged in piracy (Art 106), in the slave trade or in unauthorized broadcasting; and ships without nationality. They are of limited use in fisheries enforcement with the exception that IUU fishing vessels have been detected flying no flag and are by definition without nationality. In such cases the right of visit is exercisable by warship son the high seas.

160 The International Tribunal for the Law of the Sea (ITLOS) held in the *M/V Saiga (No. 2)* case [Saint Vincent and the Grenadines *v* Guinea, 1 July 1999, <www.itlos.org> at paragraph. 146] that "the conditions for the exercise of hot pursuit are cumulative; each of them has to be satisfied for the pursuit to be legitimate under the Convention."

161 Per David Carter, The ABC, 'The 7.30 Report', (19 August 2003).

162 Molenaar, 'Multilateral Hot Pursuit and Illegal Fishing in the Southern Ocean' above, 21.

8.11 Diplomatic Action (and sharing resources)

In late 1998, Australia and France sent a joint demarche to Mauritius to address the role of port states in assisting to eliminate illegal fishing for the Toothfish.[171] This visit was prompted by the docking of the *Salvora,* a pirate fishing vessel, pursued by Greenpeace from Australian waters adjacent to the Heard and McDonald Islands. Australian officials requested that the *Salvora* not be allowed into port to unload her catch of Toothfish without a full investigation of her fishing activities.[172] Officials in Mauritius agreed to intervene if the *Salvora* arrived at the port of Saint Loius.[173]

In 1997, the *CCAMLR* contracting parties decided to invite both Namibia and Mauritius to attend the 1998 *CCAMLR* Commission meeting as observers.[174] As both States were known port States with little regulation of the Toothfish entering its borders, it was a vital step in closing doors to the illegal trade. This followed joint diplomatic pressure by France, Australia, New Zealand and South Africa on Mauritius and Namibia to close their ports to the landing and processing of illegal catches of Toothfish.[175] As discussed in Chapter 5, both Mauritius and Namibia are *CCAMLR* parties.

Australia has also provided training to authorities in Mauritius and Namibia to assist in reaching compliance with the CDS implemented by *CCAMLR* in 1999.[176] Australia continues to send officers to monitor the landings of Toothfish at Port Louis, Mauritius, by licensed Australian fishing vessels and will do so until satisfied Mauritius has the capability to implement comprehensive monitoring and document validation.[177] In 2004 the Australia Fisheries Minister reported that Mauritius remained committed to closing the port of Saint Loius to IUU fishers.[178]

In 2002 both Australia and South Africa made representations to authorities in Mozambique urging them to prevent the unloading of Toothfish from two Uruguayan flagged vessels, the *Dorita* and the *Arvisa 1*.[179] It seems that despite diplomatic approaches to the flag State to take action in relation to false information provided

---

171 Greenpeace Press Release: 'Greenpeace welcomes Mauritius' Fisheries Minister's statement that Mauritius will join the fight against pirate fishing' 14 March 1999. <www.greenpeace.org>.

172 Ibid.

173 'Greenpeace chases pirate fishers in waters off Australia' 3 March 1999 and '13 day hot pursuit', 14 March 1999. <www.greenpeace.org>.

174 *CCAMLR*-XVI (1997) Report of the Sixteenth Meeting of the Commission, paragraph 5.36.

175 Africa News Online, *South Africa Toothfish laundering* 20 April 1998.

176 Personal communication with Mr John Davis, AFMA, 4 June 2001.

177 Report of Member's Activities in the Convention Area 2001-2002 - Australia. *CCAMLR*-XXI (2002).

178 I. Macdonald, 'Mauritius vows to close its port to illegal fishers' 27 January 2004. <www.mffc.gov.au.releases/2004>. (25 February 2004).

179 Senate Hansard, 10 December 2002, above n.132. See, also, Report of Member's Activities in the Convention Area 2001-2002 - Australia. *CCAMLR*-XXI (2002). The *Arvisa 1* has also been named the *Camouco* and the *Eternal.*

by the two vessels, Uruguay declined to withdraw the validated Catch Documents and the catch entered the international markets.[180] Australian authorities then sought the assistance of known market States urging them to hold imported shipments from either vessel until the matter was investigated. Over 200 tonnes of Toothfish product was traced to Japan, China, Hong Kong and the USA Despite the provision of detailed reports by Australia, all importing States, save the USA, declined to hold the shipments.[181] The USA seized and declared as illegal a 32 tonne shipment traced to the *Arvisa 1* and scheduled for import into America.[182]

The *Dorita* had previously been the subject of diplomatic discussions between Australia, South Africa and Mauritius. In 1999 the vessel was refused entry to South African ports on the basis that Catch Documents produced were inadequate in supporting claims of a legal high seas catch. The *Dorita* then set sail for Mauritius and following representations by Australian officials, Mauritius requested Catch Documents consistent with the *CCAMLR* CDS. The documents produced by the vessel's Master were deemed suitable by Mauritian officials.[183]

8.12 Control of nationals and flag vessels

It has been stated that if flag State control was effectively exercised then IUU fishing would not have developed into such a serious problem. The same can be said of State control over nationals. The *Compliance Agreement*, *Code of Conduct*, *UN Fish Stocks Agreement* and *IPOA-IUU* all emphasise flag State duties in relation to fishing vessels.[184] The *IPOA-IUU* also calls for all States to exercise control over their nationals.[185]

The Australian government has taken steps to strengthen control over nationals. Australian nationals involved in fishing activities in specified areas outside the AFZ 'including the Antarctic waters within 200 miles of the Australian Antarctic Territory baselines, specified areas of the high seas in relation to fishing for southern

---

180 Ibid, and see, Report of Member's Activities in the Convention Area 2001-2002 - Australia. *CCAMLR*-XXI (2002). The Australian research vessel, *Aurora Australis* sighted the two vessels within the Convention Area and on making contact was informed, falsely, that they were the Ghanian flagged *Nova Tuna 1* and the Muaritanian flagged *Kambott*.

181 Ibid.

182 Ibid. See, also, Parliamentary Secretary to the Minister for the Environment and Heritage, Media Release: 'Major blow to illegal fishing' 11 June 2002.

183 This incident was discussed at *CCAMLR*-XXI (2002). Annex 5, Report of standing Committee on Observation and Inspection, paragraph 5.6. Japan, a major market State also reported that the cargo on board the *Dorita* was accepted in full accordance with the Catch Documentation requirements and was based on vessel monitoring data provided by the flag State.

184 See generally Chapter 4.

185 *IPOA-IUU*, paragraphs 18-19.

bluefin tuna or northern bluefin tuna, *CCAMLR* sub-area 58.5.2 and other specified *CCAMLR* waters' fall under the jurisdiction of the *Fisheries Management Act.*[186]

With particular reference to the *CCAMLR* Convention Area, all Australian nationals (and vessels) are required to be licensed to fish within the *CCAMLR* Convention Area under the *Antarctic Marine Living Resources Conservation Act* 1981. In 1999 specific provisions regulating the activities of Australian flagged fishing vessels operating beyond the EEZ were introduced.[187] It is now an offence to intentionally fish on the high seas without authorisation via either a fishing concession or scientific permit.[188] It is also an offence to be equipped to fish on the high seas without authorisation.[189] Australian flagged vessels operating in waters of third States are also subject to legislative control.[190]

The activities of nationals fishing within the AFZ are also regulated. It is an offence under section 95(1)(a) of the *Fisheries Management Act* to engage in commercial fishing within the AFZ unless in possession of a fishing concession or scientific permits, that is in force authorising commercial fishing at that place. A fishing concession is defined in section 4 to include: (a) a statutory fishing right; (b) a fishing permit; (c) a foreign fishing licence. Fishing permits are granted under section 32 of the *Fisheries Management Act* whilst statutory fishing rights arise under Part 3, Division 4 of the Act.

8.13 Records of fishing vessels

The Shipping Registration Act 1981 (Cth) governs the registration of ships and the particulars relating to ships which are to be entered on the Australian Register of Ships are prescribed by regulation. The requirements for registration do not allow for foreign owned or controlled fishing vessels to be registered and flagged by Australia.[191] Whilst every Australian owned ship is required to be registered,[192] Australian owned fishing vessels are exempt from compulsory registration under the Act.[193] However the *Shipping Registration Act* also provides that an unregistered

[186] Regulations 4,4AA, 4AB, 4AC and 4AD of the *Fisheries Management Act 1991* (Cth).

[187] *Fisheries Legislation Amendment Act* (No.1) 1999 (Cth). A new Division 5A was inserted into Part 6 of the principal Act.

[188] *Fisheries Management Act 1991* (Cth) Section 105A.

[189] Ibid, Section 105B.

[190] Ibid, Section 105C.

[191] This is evident from an examination of the provisions of the *Shipping Registration Act 1981* (Cth) and the *Fisheries Management Act 1991* (Cth). See also AUS-NPOA-IUU, 22.

[192] *Shipping Registration Act* 1981 *(Cth)* section 12. The term 'Australian owned ship' is defined in section 8 and includes ships owned by two or more persons where the majority of those persons are Australian nationals.

[193] Ibid. Section 13. However they may be registered. Section 14.

Australian owned fishing vessel shall not depart from an Australian port to a place outside Australia, that is, beyond the limits of the territorial sea.[194] In practical terms this means Australian owned fishing vessels must be registered to fish beyond the territorial sea.[195]

The information required to be recorded on all registered ships whether or not authorised to fish on the high seas, reflects the requirements listed in the *Compliance Agreement*. Table 8.5 lists the information required under Australian law to be entered on to the Australian Register of Ships.[196] The information or detail required is similar to that required under the *Compliance Agreement*. The Australian Register of Ships may be inspected on payment of the prescribed fee.[197]

[194] Ibid, Section 68.

[195] See also the AUS-NPOA-IUU, 22.

[196] The information in Table 8.5 is obtained from the AUS-NPOA-IUU, 23.

[197] Shipping Registration Act 1981 (Cth) Section 57.

The economic benefits of providing port access which would be lost to port States on the implementation of the CDS (or similar regional measures), will continue to play a decisive role in the extension of port controls regionally. Port States, with existing landing and transhipment infrastructure will require an economic or political inducement to alter their practices. There are considerable profits to be realised in providing port facilities to fishing vessels and particularly IUU vessels. Further, as demonstrated in the review of *CCAMLR* measures, port State controls can be undermined by just one conveniently located port at which the CDS is not implemented.

Finally, the presumption of non-compliance as raised in the *IPOA-IUU* can be implemented. The *CCAMLR* CDS is supported by a number of additional conservation measures[12] including CM10-07 which operates to create a presumption of noncompliance with respect to the fishing vessels of non-contracting parties *sighted engaged* in fishing within the Convention Area. The conservation measure requires all contracting parties to inspect such vessels upon their entry into port and to prohibit any transhipments or landings until the vessel's crew have established the fish on board have been caught either outside the Convention area or in accordance with conservation measures.

The *IPOA-IUU* also addresses this aspect of port State controls however suggests that the presumption of non-compliance be triggered when the non-contracting vessel is *identified* as having engaged in IUU fishing.[13] The *IPOA-IUU* trigger is a simpler one to establish than the *CCAMLR* requirement that non-contracting party vessels be *sighted engaged* in fishing before the presumption of non-compliance can be raised. For example, support vessels would fall under the *IPOA-IUU* test, yet escape identification for the purposes of *CCAMLR* conservation measure 10-07.

The fluid nature of corporate IUU fishing, a factor addressed in Chapter 2, has resulted in an increase in use of re-supply and trans-shipment vessels. The benefit of such support vessels to the IUU operator is that they may avoid port and market State controls. By transshipping catches, the IUU vessel can remain *in situ* and the support vessel can land the IUU fish. Through transshipping the reported origin of Toothfish may be falsified on CDS documents. Alternatively, IUU catches may be mixed with legally caught fish with the more valuable whole fish being sold under legal documentation whilst less valuable fish can be sold into markets where port States controls are not so strictly applied.[14] Under the *IPOA-IUU* test, these support vessels would be subject to the presumption of noncompliance. The Australian legislation adopts the *IPOA-IUU* approach, even though the *Fisheries Management Act* offences were enacted in 1991, a decade before the FAO adoption of the *IPOA-*

[12] See Chapter 6.

[13] *IPOA-IUU*, paragraph 63.

[14] This fact is supported through the submissions of ASOC to *CCAMLR*-XXIII(2004). See *CCAMLR*-XXIII/BG/27 Rev.2 (4 November 2004). Document accessed with the permission of ASOC, copy on file.

IUU. Under Australian law, support vessels are subject to significant fines and subject to forfeiture under the statutory forfeiture regime in s 106A of the *Fisheries Management Act.*

Long Term Improvement Strategies

The implementation of regional port State MoUs, modelled on the structure created by the IMO port State MoUs for ship safety and environmental controls, should be investigated further. Regional MoUs have a number of possible advantages including, supporting existing RFMOs through the implementation of uniform and complementary controls, supporting port States considering implementing port controls and providing coverage of areas of high seas presently lacking regional regulation. The HSTF has recommended consideration be given to synergies between RFMOs.[15]

It is likely that regionally applied MoUs establishing minimum requirements for port State inspections would reduce the possibility of some ports remaining outside the port State control schemes implemented by RFMOs thereby becoming attractive ports for IUU fishers. This is based on the argument that the reduction in competition amongst ports resulting from the IMO safety and environmental controls for shipping was an important element to successful port State controls.[16] Similarly, regional MoUs for port State inspections would level the playing field amongst port States and possibly reduce the likelihood of port States operating in support of the IUU fishing industry. To further support the successful implementation of port State controls, the provision of assistance and training to developing States seeking to apply port controls(such as that given to Mauritius by Australia) should be made a priority.

As discussed in Part 3, Chapter 4, the FAO has investigated the introduction of MoUs on port State controls.[17] Concerns were raised at the FAO Technical Consultation to Review Port State Measures to Combat Illegal, Unreported and Unregulated Fishing[18] that using the MoU to implement the aims of the IPOA-IUU would introduce a further bureaucratic layer[19] and potentially delay progress. Ultimately the consensus at the Technical consultation was to adopt a Model Scheme on Port State Measures to combat Illegal, Unreported and Unregulated Fishing.[20]

---

15 Lodge and Meere, above, 6-7. The authors recommend exploring ways of encouraging cooperation between the main tuna RFMOs and suggest shared vessel registers, common catch documentation schemes and the review of trade and finance flows.

16 T.L. McDorman, 'Regional Port State Control Agreements: Some Issues of International Law' (2000) 5 *Ocean and Coastal Law Journal* 207, 209.

17 See Chapter 3, n 112.

18 Rome, 31 August -2 September 2004.

19 FAO Fisheries Report, *Report of the Technical Consultation to review Port State Measures to combat Illegal, Unreported and Unregulated Fishing* (2004) N0. 759, paragraphs 11-12.

20 Ibid, Appendix E.

IUU Vessel or FOC Registers

As evidenced in the review of the *CCAMLR* IUU Vessel Lists in Chapter 6, clear and objective criteria for listing vessels must be established. The *IPOA-IUU* makes reference to the development and maintenance of records of vessels fishing in the area of application of the RFMO including those authorised to fish and those engaged in or supporting IUU fishing. In this regard the *CCAMLR* Commission introduced two IUU Vessel Lists for member and non member States whose flagged fishing vessels engage in IUU fishing.

The detailed discussions during the 2003 and 2004 *CCAMLR* Commission meetings about the inclusion of vessels on the IUU Vessel List for member States, indicates the potential impact of such lists. Quite apart form the negative publicity associated with listing on the IUU Vessel List, the onus of establishing that any Toothfish on board listed vessels has been caught in accordance with *CCAMLR* conservation measures, rests firmly upon the vessel and the Master. With measures to reduce opportunities for submitting fraudulently completely CDS documents being considered and trialled by *CCAMLR*, IUU vessels will find it increasingly difficult to land their fish and when seeking to land their fish at *CCAMLR* contracting party ports, face the task of disproving the presumption of non-compliance.

Licensed Vessel List

The decision of the *CCAMLR* Commission in 2004 to place on the secure section of the *CCAMLR* website a list of all vessels licensed to fish within the Convention Area together with details of the vessel's make, identifying flag and radio call sign, ownership details, license holder information and colour photographs of the vessel, will support the deterrence of IUU fishing. *CCAMLR* members can refer to the list of licensed vessels and use the photographs and detailed information provided to determine whether any fishing vessel found within the Convention Area is licensed to fish. The presumption being if it is not on the *CCAMLR* licensed vessel, the vessel is engaging in IUU fishing. Given that many IUU vessels are not clearly marked or identifiable, access to what has been termed a 'white' list will enable *CCAMLR* contracting parties to quickly verify the status of sighted vessels.

9.3 Regional cooperation

Improved fisheries management through regional cooperation is a consistent theme in the international instruments reviewed in Part 3, Chapter 4. The Report of the HSTF on its 9 March 2005 meeting reflects this approach.[32] The HSTF

32 Lodge and Meere, 'High Seas Governance' above, 5-6.

recommended the promotion of better coordination between RFMOs and increased harmonisation of measures.[33]

The *CCAMLR* Commission is a successful example of regional cooperation to deter IUU fishing. The implementation of the CDS and encouragement of non-contracting parties to participate has resulted in reports of non CDS fish selling at lower prices. Three non-contracting parties now implement the *CCAMLR* CDS[34] and two former port States, Mauritius and Namibia have become *CCAMLR* parties.[35]

The *CCAMLR* Commission has also looked outside the Convention Area to cooperate with other RFMOs. SEAFC a newly formed RFMO shares a boundary with *CCAMLR* and regular communication with SEAFC members will be required. Further support of *CCAMLR*'s regional cooperation can be seen in the Commission's annual meeting Reports in which linkages with numerous RFMOs are noted.

As a coastal State and RFMO member, Australia has been active in regional cooperation through membership of the *CCAMLR* Commission, participation in the *CCAMLR* E-CDS and E-VMS trials, supporting neighbouring regional States in deterring IUU fishing, forming a bilateral treaty for cooperative surveillance with France and entering negotiations with South Africa for a similar treaty and supporting those NGOs actively campaigning against IUU fishing.

9.3.1 National Measures

National measures have been successful in deterring IUU fishing when implemented by the coastal State within ports and the EEZ for they are based on coastal State sovereignty and are not restricted by the international principles of the freedom of fishing the high seas, flag State sovereignty or *pacta tertii*. On this basis coastal states represent a powerful tool to assist RFMOs in deterring IUU fishing through their support of RFMO measures. As discussed in Chapter 5, Australia supported the work of the *CCAMLR* by implementing conservation measures within the Heard and MacDonald Island's fishing zone and applying the *CCAMLR* CDS and C-VMS conservation measures.

The Australian government has conducted considerable legislative and policy reviews and implemented significant amendments to the *Fisheries Management Act* to deter IUU fishing. Fines have been increased, provisions for vessel forfeiture introduced and the policy indicates a commitment to continued surveillance and cooperation with other States. Whilst these developments are encouraging and provide a model for other coastal States, care should be taken to ensure the measures comply with the rule of law.

---

33 Ibid.

34 The Peoples Republic of China. The Seychelles and Singapore. See Chapter 6.

35 Ibid. Mauritius is an acceding States and applies the CDS. Namibia joined the Commission in 2001.

9.4 Conclusion

The aim of this study has been to demonstrate, notwithstanding the many factors (legal and otherwise) contributing to or 'driving' the development of IUU fishing, that there are measures, based upon international law, which RFMOs and coastal States can implement to deter IUU fishing. It has been demonstrated that the measures implemented by *CCAMLR*, as an RFMO and Australia, as a coastal State, take account of the relevant international instruments addressing IUU fishing and present as a working model of measures to deter IUU fishing in the Southern Ocean.

Whilst there have been early indications of success as evidenced by the changed behaviour of IUU operators, it is too soon to conduct a considered review of their impact. It is concluded that whilst the early signs are promising, before their full impact can be assessed more time must pass, during which all the measures should continue to be consistently and uniformly applied.

Treaties, Agreements and Declarations

1947 General Agreement on Tariffs and Trade 55 UNTS 194

1949 International Convention for the Northwest Atlantic Fisheries 157 UNTS 158

1958 Convention on the High Seas 250 UNTS 82

1958 Convention on Fishing and Conservation of the Living Resources of the High Seas 559 UNTS 285

1958 Convention on the Territorial Sea and the Contiguous Zone 516 UNTS 205

1959 Antarctic Treaty 402 UNTS 71

1969 Vienna Convention on the Law of Treaties 155 UNTS 331

1973 Convention on International Trade in Endangered Species 12 ILM 1085

1978 Convention on Future Multilateral Cooperation in the Northwest Atlantic Fisheries 1135 UNTS 369

1980 Convention for the Conservation of Antarctic Marine Living Resources 19 ILM 841

1982 United Nations Convention on the Law of the Sea 1833 UNTS 31363

1986 United Nations Convention on the Conditions for Registration of Ships 1986

1988 Convention on the Regulation of Antarctic Mineral Resource Activities 27 ILM 859

1991 Protocol to the Antarctic Treaty on Environmental Protection 30 ILM 1461

1992 Cancun Declaration on Responsible Fishing

1992 International Convention for the Conservation of Atlantic Tuna 32 ILM 1232

1993 Agreement to Promote Compliance with International Conservation and Management Measures by Fishing Vessels on the High Seas 33 ILM 968

1995 United Nation Agreement for the Implementation of the Provisions of the United Nations Convention of the Law of the Sea of 10 December 1982 relating to the Conservation and Management of Straddling Fish Stocks and Highly Migratory Fish Stocks 24 ILM 1542

1995 Code of Conduct for Responsible Fisheries

1999 Rome Declaration on the Implementation of the Code of Conduct for Responsible Fisheries

2001 International Plan of Action- Illegal, Unreported and Unregulated Fishing

2003 Treaty between the Government of Australia and the Government of the French Republic on cooperation in the maritime areas adjacent to the French Southern and Antarctic Territories (TAAF), Heard and McDonald Islands 2003 ATNIF 20

Legislation

Australian legislation

The Constitution of Australia Act 1900

Seas and Submerged Lands Act 1973

Antarctic Marine Living Resources Conservation Act 1981

Shipping Registration Act 1981

Heard and MacDonald Islands' Environmental Protection and Management Ordinance 1987

Fisheries Administration Act 1991.

Fisheries Management Act 1991.

Maritime Legislation Amendment Act 1994.

Fisheries Legislation Amendment Act 1999.

Fisheries Legislation Amendment (High Seas Fishing Activities and Other Matters) Act 2004

List of Cases

Australian Cases

Chow Hung Ching v. The King (1949) 77 CLR 449

Chiou Yaou Fa v Morris (1987) FLR 36

Gayfer v Bere (1998) 102 A Crim R. 208.

R v Paz and Santome (unreported decisions of the District Court of Western Australia, 9 October 1998 and 14 October 1998)

Readhead and Ors v Admiralty Marshall, Western Australia District Registry and Ors (1998) 87 FCR 229

Bergensbanken Asa v The Ship 'Aliza Galcia' [1998] 1322 FCA

Bergensbanken Asa v The Ship 'Aliza Galcial' [1998] 1642 FCA

Mackry v Diman [1999] NTSC 96 (unreported 7 September 1999, Riley J)

Olbers v Commonwealth of Australia (No 3) [2003] FCA 651

Olbers Co Ltd v Commonwealth of Australia [2004] FCAFC 262

Perez v R (1999) 21 WAR 470

International Tribunal for the Law of the Sea

St Vincent and the Grenadines v Guinea (The M/V Saiga Case) ITLOS Case No. 4 (1997)

St Vincent and the Grenadines v Guinea (The M/V Saiga (No.2)Case) (Merits) ITLOS Case No.8 (1999)

Australia v. Japan and New Zealand v. Japan (Southern Bluefin Tuna Case) ITLOS Cases No. 3 and 4 (1999)

Panama v France (The Camouco Case) ITLOS Case No. 5 (2000)

Seychelles v France (The Monte Confurco Case) ITLOS Case No. 6 (2000)

Belize v France (The Grand Prince Case) ITLOS Case No. 8 Judgment dated 20 (2001)

Russian Federation v Australia (The Volga Case) ITLOS Case No. 11 (2002)

Malaysia v Singapore ILTOS Case No. 12 (Order) (2003)

International Court of Justice (and other International Courts)

Great Britain v France, Decision of the Permanent Court of Arbitration 8 August 1905 The decision of the Court of Arbitration is reproduced in (1908) 2 American Journal of International Law 921-928

The Case of the S.S. Lotus (1927) P.C.I.J. Series A No.10, 8

Netherlands v United States of America (1928) Permanent Court of Arbitration, 2 RIAA 829

Austro-German Customs Union Case Advisory opinion of the Permanent Court of International Justice (1931)

Anglo-Norwegian Fisheries Case (1951) ICJ Reports 116

Rights of Nationals of the United States of America v Morocco (1952) ICJ Reports 212

Liechtenstein v Guatemala (1955) ICJ Reports. 4

Advisory Opinion on the Constitution of the Maritime Safety Committee of the Inter-Governmental Maritime Consultative Organisation [1960] ICJ Reports, 150

North Sea Continental Shelf Case (1969) ICJ Reports 51

United Kingdom v Iceland and Germany v Iceland (1974) ICJ Reports 3 and174

Case concerning Military and Paramilitary Activities in and against Nicaragua (Nicaragua v United States) (1986) ICJ Report 14

Hungary v Slovakia (1997) ICJ Reports 7

Shrimp Turtle WTO Appellate Body Report on United States import prohibition on certain Shrimp and Shrimp Products 38 ILM 118 (12 October 1998).

Legality of the Threat or Use of Nuclear Weapons (1996) ICJ Rep

Other Jurisdictions

Young v Hichens (1844) 6 QB 606.

State of Ohio v Shaw (1902) 65 NE 875

Mariana Flora 24 U.S. (11Wheat) 1 (1826)

Naim-Molvan v Attorney-General for Palestine 1948 AC 351

Bibliography

1. Books

Auburn, F.M. Antarctic Law and Politics (C.Hurst and Company 1982)

Birnie P. and Boyle, A. International Law and the Environment (Clarendon Press 2nd ed, 2002)

Boczek, B.A. Flags of Convenience –An International legal study (Harvard University Press 1962)

Boggs, S.W. The Polar Regions: Geographical and historical data for consideration in a study of claims to sovereignty in the Artic and Antarctic regions (William S.Hein and Co.1933, 1990)

Brownlie, I. Principles of Public International Law (Oxford University Press 6th ed 2003)

Burke, W.T. The New International Law of Fisheries (UNCLOS 1982 and Beyond) (Clarendon Press 1994)

Carlisle, R. Sovereignty for Sale (Naval Institute Press 1981)

Caron, D. 'Ships, Nationality and Status' Encyclopaedia of Public International Law Vol. 4 (2000)

Caron, D. and Scheiber, H. Bringing New Law to Ocean Waters (Martinus Nijhoff 2004)

Christy, Jr. Francis T. and Scott. A, The Common Wealth in Ocean Fisheries (John Hopkins Press 1965).

Churchill, R. and Lowe, A, The Law of the Sea (Manchester University Press 3rd ed, 1988)

Clingan, T. The Law of the Sea: Ocean law and Policy (Austin & Winfield 1994)

Colombus, J. The International Law of the Sea (D.McKay Co. 1976)

Cushing, D. The Provident Sea (Cambridge University Press 1988)

Dekker, I.F. and Post, H.G. (eds) On the Foundations and Sources of International Law (T.M.C. Asser Press 2003)

Elder, J. The Royal Fisheries Companies of the 17th Century (James MacLehose and Sons 1912)

Elferink, A.O. and Rothwell, D.R. (eds) The Law of the Sea and Polar Maritime Delimitation and Jurisdiction (Kluwer Law International 2001)

Elferink, A.O. and Rothwell, D.R. (eds) Oceans Management in the 21st Century (Kluwer Law International 2004).

Fogg, G. A History of Antarctic Science (Cambridge University Press 1992)

Baird, R. 'Testing the Waters: Fine tuning the provisions of the Fisheries Management Act 1991 applicable to foreign fishing boats' (2004) 32 University of Western Australia Law Review 63.

Baird, R. 'Corporate Criminals and their Involvement in IUU Fishing: an Australian Perspective' (2005) International Fisheries Law and Policy Review (in print)

Balton, D. 'Strengthening the Law of the Sea: The New Agreement on Straddling Fish Stocks and Highly Migratory Fish Stocks' (1996) 27 Ocean Development and International Law 125

Balton, D. 'The Bering Sea Doughnut Hole Convention: Regional Solution, Global Implications' in Stokke, O (ed), Governing High Seas Fisheries: The Interplay of Global and Regional Rules (2001) 143

Barrett, M. 'Illegal Fishing Zones subject to National Jurisdiction' (1998) 5 James Cook University Law Review 1

Barston, R. 'The Law of the Sea and Regional Fisheries Organisations' (1999) 14(3) International Journal of Marine and Coastal Law 333

Bateman, S. and Rothwell, D. (eds) Southern Ocean Fishing: Policy Challenges for Australia Centre for Maritime Policy (Maritime Policy paper No.7) (1998)

Bayitch, S.A. 'International Fishery Problems in the Western Hemisphere' (1955-56) 10 Miami Law Quarterly 499

Behnam, A. 'Ending Flag State Control' in A.Kirchner, International Maritime Environmental Law (2003) 123

Botsford, L.W., Castills, J.C. and Peterson, C.H., 'The Management of Fisheries and Marine Ecosystems' (1997) 277 (5325) Science 507

Brown, H. 'The United Nations Conference on Straddling Fish Stocks and Highly Migratory Stocks: An Analysis of International Environmental Law and the Conference's Final Agreement' (1996) Vermont Law Review 547

Burke, W.T. ′Fishing in the Bering Sea Donut Hole: Straddling Stocks and the New International Law of Fisheries" (1989) 16 Ecology Law Quarterly 285

Campbell, B. 'Maritime Legislation Amendment Act 1994' (1994) 5 (2) Public Law Review 141

Caron, D. 'Ships, Nationality and Status' Encyclopedia of Public International Law Vol. 4 (2000) 400

Carroz, J.E. 'International Aspects of Fisheries Management under the New Regime of the Oceans' (1984) 21(3) San Diego Law Review 513

Charlesworth, H. 'Customary International Law and the Nicaragua Case' (1991) 11 Australian Yearbook of International Law 1

Christopherson, M. 'Towards a rational harvest: The United Nations Agreement on Straddling Fish Stocks and Highly Migratory Species' (1996) 5 Minnesota Journal of Global Trade 357

Christy Jr. F.T. 'Marine Resources and the Freedom of the Seas' (1968) 8 Natural Resources Journal 424

Clarke, B. and Hemmings, A. 'Problems and Prospects for the Convention on the Conservation of Antarctic Marine Living Resources Twenty Years On' (2001) 4 Journal of International Wildlife Law and Policy 47

Collett, M. 'Achieving effective International Fisheries Management- A Critical Analysis of the United Nations Conference on Straddling Fish Stocks' (1995) 4 Dalhousie Journal of Legal Studies 1

Costanza, R. et al. ' Principles for the Sustainable Governance of the Oceans' (1996) 281 (5374) Science 198

Daggett, A.P. 'The Regulation of Maritime Fisheries by Treaty' (1934) 28 American Journal of International Law 693

de la Fayette, L. 'Access to Ports in International Law' (1996)11 International Journal of Maritime and Coastal Law 1

Davies, P. 'The EC/Canadian Fisheries Dispute in the North west Atlantic' (1995) 44 (4) International and Comparative Law Quarterly 927

Davies P. and Redgwell, C. 'The International Legal Regulation of Straddling Fish Stocks' (1997) 67 British Yearbook of International Law 199

Edeson, W. 'The Code of Conduct for Responsible Fisheries' (1996) 11 International Journal of Marine and Coastal Law 233

Edeson, W. 'Closing the Gap: The Role of 'Soft' International Instruments to Control Fishing' (2000) Australian Yearbook of International Law 83

Elferink, A.O.G, 'The Genuine Link Concept: Time for a Post Mortem?' in I.F. Dekker and H.G. Post (eds) On the Foundations and Sources of International Law (2003) 41.

Fenn, P.T., 'Justinian and the Freedom of the Sea' (1925) 19 American Journal of International Law 717

Freestone, D. 'The Effective Conservation and Management of High Seas Living Resources – Towards a New Regime' (1994) 5 Canterbury Law Review 341

Freestone, D. and Makuch, Z., 'The 1995 UN Straddling Stocks Agreement' (1996) 7 Yearbook of International Environmental Law 3.

Gillmore, W.C. 'Hot Pursuit: The case of R v Mills and Others' (1995) 44(4) International and Comparative Law Quarterly 949

Gordon, H. 'The Economic Theory of a Common Property Resource: The Fishery' (1954) 62 Journal of Political Economy 124

Green, J. and Agnew D. 'Catch Documentation Schemes to combat Illegal, Unreported and Unregulated fishing: CCAMLR's experience with the Southern Ocean Toothfish' (2002) 16 Ocean Yearbook 171

Grynberg, R. 'WTO fisheries subsidies negotiations: implications for fisheries access arrangements and sustainable management' (2003) 27 Marine Policy 499-511

3. Australian Governmental Material

3.1 Parliamentary papers

Commonwealth, House of Representatives Standing Committee on Primary Industries, Resources and Rural and Regional Affairs Managing Commonwealth Fisheries: The Last Frontier (1997)

Commonwealth Policy Statement, Department of Primary Industries and Energy, New Directions for Commonwealth Fisheries Management in the 1990s (1997)

Australia's Strategic Policy (1997)

Commonwealth, House of Representatives Standing Committee on Legal and Constitutional Affairs, Australian law in Antarctica (1992)

Australian National Audit Office, Audit Report No 32, Commonwealth Fisheries Management – Australian Fisheries Management Authority Vols. 1 & 2 (1995-96)

3.2 Departmental Papers

Department of Foreign Affairs and Trade, Report of the Australian Delegation, Eleventh Session of the third United Nation's Conference on the Law of the Sea (1982).

Department of Agriculture, Fisheries and Forestry, Australia's Acceptance of the FAO Compliance Agreement (November 2000)

3.3 Joint Standing Committee on Treaties

Joint Standing Committee on Treaties Report No 47 August 2002.

3.4 Media Releases/Statements (in chronological order)

The Australian Antarctic Division, 'Toothfish – 10 Facts' undated

Australian Fisheries Management Authority, 'Commonwealth Fisheries - Heard and McDonalds Islands: Principle Commercial Fish Species' Fact Sheet, undated

Australian Fisheries Management Authority, 'The AFMA Southern Ocean Surveillance Program' Fact Sheet, undated

Australian Fisheries Management Authority, 'Operation Cosmo- Hot Pursuit and apprehension of suspected Illegal Fishing Vessel' (undated)

Australian Fisheries, Forestry and Conservation, 'Global Fisheries Issues affecting Australia', undated

Minister for Resources and Energy and Minister for Foreign Affairs and Trade, Press Release, 'Australia welcomes Norwegian Government action on Illegal Fishing' (22 July 1998).

Australian Fisheries Management Authority, 'Foreign Fishing Vessel Aliza Galcial sets sail' (23 December 1998).

Australian Fisheries Management Authority, 'Commonwealth Fisheries - Heard and McDonalds Islands- Principal Commercial Fish Species', Fact Sheet (2001).

Minister for Defence, 'Minister Congratulates ADF and AFMA on Successful Mission' MIN94/01 (12 April 2001).

Minister for Fisheries and Conservation, 'Recent Government responses to illegal fishing in the Southern Ocean' (7 June 2001).

Minister for Forestry, Fisheries and Conservation, 'Record Hot Pursuit Vessel berths in Fremantle Harbour' AFFA01/35TU (5 May 2001).

Australian Fisheries Management Authority, 'Operation Sutton – Apprehension of suspected illegal fishing vessels' (March/April 2002).

Minister for Environment and Heritage, 'Australia Seeks more protection for Toothfish' (5 June 2002).

Parliamentary Secretary to Minister for Environment and Heritage, 'Major Blow to illegal fishing' (11 June 2002).

Minister for Fisheries and Conservation, 'Government to get tougher on Toothfish pirates' (25 July 2002)

Minister for Fisheries and Conservation, 'Increased patrols to target pirates' (1 October 2002).

Parliamentary Secretary to Minister for Environment and Heritage, 'Australia steps up the battle on illegal fishing' (2 November 2002).

Minister for Environment and Heritage, 'Toothfish: Australia achieves commitment from CITES for greater protection and withdraws nomination' (14 November 2002)

Joint Statement, Minister for Foreign Affairs, Attorney General, and Minister for Forestry, Fisheries and Conservation, 'Australia strikes another blow against illegal fishing' (24 December 2002).

Minister for Fisheries, Forestry and Conservation, 'Australia steps up the battle on illegal fishing' (9 February 2003).

Minister for Fisheries, Forestry and Conservation, 'Australia calls on peak UN fisheries body to take action on illegal fishing' (26 February 2003)

Statement to FAO Committee on Fisheries (25th Session) 'Australia calls on peak UN Fisheries body to take action on illegal fishing' (26 February 2003).

Joint Statement by Minister for Forestry and Conservation and Minister for Justice and Customs, '$12 million budget boost to fight illegal fishing in the southern Ocean' (13 May 2003).

'Minister for Fisheries, Forestry and Conservation, 'The Howard Government's efforts to deter illegal fishing activities' Address at National Press Club, Canberra (19 August 2003).

Minister for Fisheries, Forestry and Conservation, 'Dangerous Ice pursuit of Viarsa now longest in Australian maritime history' (25 August 2003)

Minister for Fisheries, Forestry and Conservation, 'Bringing back the Viarsa' (30 August 2003)

Minister for Fisheries, Forestry and Conservation, 'Tough new penalties for illegal fishing' (26 November 2003)

Minister for Fisheries, Forestry and Conservation, 'Pirates beware, Australia front and centre in the fight against illegal fishing' (2 December 2003)

Minister for Fisheries, Forestry and Conservation, 'First illegal fishers caught for 2004' (7 January 2004)

Minister for Fisheries, Forestry and Conservation and Minister for Defence, 'Navy catches suspected Illegal Fishing Vessel' (24 January 2004)

Minister for Fisheries, Forestry and Conservation 'Mauritius vows to close its port to illegal fishers' (27 January 2004)

Minister for Fisheries, Forestry and Conservation, 'Review of catch landings in Australia' (29 January 2004)

Minister for Fisheries, Forestry and Conservation 'New Chapter in maritime law- Attempt to claim back the Volga rejected ' (13 March 2004)

Minister for Fisheries, Forestry and Conservation and Australian Minister for Justice and Customs, 'Toothfish tender nets over $2M' (23 April 2004)

Joint Media Release, Minister for Justice and Customs and Minister for Fisheries , Forestry and Conservation, 'New role for Toothfish Pirate' (21 May 2004)

Minister for Fisheries, Forestry and Conservation and C.Ellison, Minister for Justice and Customs, 'Warning to fish poachers as armed patrols begin' (19 November 2004)

Minster for Fisheries, Forestry and Conservation and C.Ellison, Minister for Justice and Customs, 'Flag of Convenience vessels flaunt international rules' (Joint Statement 4 March 2005)

Minister for Fisheries, Forestry and Conservation, 'New international resolve to tackle illegal fishing' (15 March 2005)

Department of Agriculture, Fisheries and Forestry, 'Overview of Illegal Unreported and Unregulated Fishing. (16 March 2005)

Minister for Fisheries, Forestry and Conservation, 'Long-term Commitment to Southern Ocean armed patrols' (10 May 2005)

Minister for Fisheries, Forestry and Conservation 'Ocean the new hotspot for criminals' (13 May 2005)

Minister for Fisheries, Forestry and Conservation, 'Century up in fight against illegal fishing' (12 June 2005)

3.5 Hansard

Senate Hansard, Question No. 730, 10 December 2002, 7659.

Senate Hansard 12 December 2002.

Senate Estimates, Program 1.8, International Legal Interests- Foreign Affairs and Trade Committee 26 December 1998, 136.

3.6 Departmental and Authority Annual Reports and Publications

Department of Agriculture, Fisheries and Forestry Annual Reports (1996 to 2002).

Australian Fisheries Management Authority Annual Reports (1996 to 2002).

'Fishing Future' Newsletter of the Australian Fisheries Management Authority, Vol.2. Issue 4, December 2004,

Heard and McDonald Islands' Fishery Management Plan 2002

4. Non Governmental Organisation Materials

4.1 Research Papers

Greenpeace and International Transport Federation, Troubled Waters: Fishing, Pollution and Flags of Convenience (April 1999)

ISOFISH, The Involvement of Mauritius in the trade in Patagonian Toothfish from illegal and unregulated long line fishing in the Southern Ocean and what might be done about it, Occasional Report No. 1 (August 1998, 3rd edition)

ISOFISH, Toothfish Poachers change vessel names in an attempt to avoid identification by the CCAMLR Catch Documentation Scheme Occasional Report No. 4. (May 2000)

4.2 Press Releases

ISOFISH, 'Toothfish test boat turned away' (28 June 2000)

ISOFISH, 'Australian and South African authorities arrest poacher' (14 April 2001)

ISOFISH, 'Salvora' Fact Sheet, undated

ISOFISH, 'Patagonian Toothfish Profile' undated

ISOFISH, 'Russian trawler steals fish from New Zealand waters' (25 July 2000)

ISOFISH, 'Callous Spanish Fishing Companies put 'White Gold' ahead of crew safety in the Southern Ocean' (28 October 2000)

Greenpeace, 'The case of the Salvora' undated

Greenpeace, 'Greenpeace chase pirate fishers in waters off Australia' (3 March 1999)

Greenpeace, '13 day hot pursuit'(14 March 1999)

Greenpeace, 'Greenpeace welcomes Mauritius' Fisheries Minister's Statement that Mauritius will join the fight against pirate fishing' (14 March 1999)

Greenpeace, 'Dodging the Rules: Flag of Convenience Fishing' (2000)

Greenpeace, 'Australia leads on pirate Port battle' (6 April 2000)

Greenpeace, 'Southern Ocean Pirate Fishing Expedition' (2000)

ASOC, 'An Australian fisheries patrol vessel detected and warned off a Foreign fishing vessel' (4 March 1999)

ASOC, 'Going to hell in a fishing basket' (29 October 1999)

ASOC, 'Conservationists congratulate Governments for taking on the Southern Ocean Toothfish Poachers' (5 November 1999)

ASOC, 'Environmentalists and Trade Union Leaders have called for urgent action by Governments to end pirate fishing in the World's Oceans' (1999)

ASOC, 'Time for Toothfish Moratorium' (23 October 2000)

5. United Nations Material

5.1 Resolutions and Reports

UN General Assembly Resolution, 899(IX) (1954)

UN General Assembly Resolution, 37/209, 20 December 1982

UN General Assembly Resolution, 22 December 1992

UN General Assembly Resolution, 10 December 1995

UN General Assembly Resolution, 54/32 (1999)

UN General Assembly Resolution, 54/32 (2000)

UN General Assembly Resolution, 39/25, 17 January 2005

'Memorandum on the regime of the high seas', prepared by the UN Secretariat. UN Doc A/CN.4/32 Yearbook of the International Law Commission (1950) Vol. II.

Memorandum on the Regime of the High Seas – Replies from Governments to Questionnaires of the International Law Commission, UN Doc A/CN.4/19 Yearbook of the International Law Commission (1950) Vol. II.

Report of the International Law Commission (ILC) to the General Assembly 1956, Yearbook of the ILC (1956)

Report of the International Law Commission to the United Nations General Assembly 1955, Yearbook of the ILC (1955) Vol. II

Report of the UN Conference on the Environment and Development (UNCED) Agenda 21, Chapter17. (1992)

Division for Ocean Affairs and the Law of the Sea, 'The United Nations Convention on the Law of the Sea- a Historical Perspective' (1998)

Report of the UN Secretary General on Oceans and the Law of the Sea to the 53th session of the UN General Assembly. UNDOC A/54/429 (30 September 1999)

Report of the UN Secretary General on Oceans and the Law of the Sea to the 54th session of the UN General Assembly. UNDOC A/55/61 (20 March 2000)

5.2 Press Releases and Statements

Statement at opening of the Organisational Session by the Chairman of the UN Fish Stocks Conference, 19 April 1993 (A/CONF.164.7 4 May 1993)

Statement made by Chairman of the UN Fish Stocks Conference upon the adoption of the UN Fish Stocks Agreement 20 September 1995 (A/CONF.164/35)

6. FAO

6.1 Papers

Garcia, S.M. 'World Review of highly migratory species and straddling stocks' FAO Technical Paper No. 337 (1994)

Elferink, A.G. 'The impact of Article 7(2) of the Fish Stocks Agreement on the Formulation of Conservation and Management Measures for Straddling and Highly Migratory Fish Stocks' FAO Legal Papers Online No. 4 (August 1999)

Lugten, G. 'A Review of measures taken by Regional Fisheries Bodies to address Contemporary Issues' FAO Fisheries Circular No. 940 (1999)

Franckx, E. 'Pacta Tertiis and the Agreement for the Implementation of the Provisions of the United Nations Convention on the Law of the Sea of December 1982 Relating to the Conservation and Management of Straddling Fish Stocks and Highly Migratory Fish Stocks' FAO Legal Papers Online No.8 (June 2000)

Doulman, D., 'IUU Fishing - A mandate for an International Plan of Action' (paper presented to the Expert Consultation on IUU Fishing, Sydney, Australia, 15-19 May 2000)

D.J.Doulman, 'A General overview of some aspects of Illegal, Unreported and Unregulated Fishing' April 2000 (paper presented to the Expert Consultation on IUU Fishing, Sydney, Australia, 15-19 May 2000)

Edeson, W. 'Tools to address IUU Fishing – The current legal situation' (paper presented to the Expert Consultation on IUU Fishing, Sydney, Australia, 15-19 May 2000)

Lobach, T. 'Port State Controls of Foreign Fishing Vessels' FAO Legal Papers Online No.29 (May 2002)

Swan, J. 'Fishing vessels operating under Open Registers and the Exercise of Flag State Responsibilities- Information and Options' FAO Fisheries Circular No. 980 (2002)

'Port State controls of foreign fishing vessels' FAO Fisheries Circular No. 987 (2003).

9. Conferences

Canadian Department of Fisheries and Oceans, 'Illegal, Unreported and Unregulated (IUU) Fishing and International Fisheries Governance Key Topics at International Meeting' Media Release 11 March 2005. <www.dfo-mpo.gc.ca/media/newsrel/2005/hq-ac08_e.htm> (20 March 2004)

International Conference on the Governance of High Seas Fisheries and the UN Fish Stocks Agreement – Moving from Words to Action, St John's, Newfoundland and Labrador, Canada, 1-5 May 2005.

10. Newspapers

10.1 Australian

AP/AAP 'Opening round of battle for Toothfish in Antarctica' 30 April 1997.

AFP/AP 001PAC 'Nations discuss polar surveillance fleet for fish pirates' 8 May 1997.

AAP Press Release 'Valuable Fish Species being plundered in the Southern Ocean' 18 May 1997.

AAP, 'The Australian A.P. Reports on ISOFISH' 31 March 1999.

ABC News Online 'Charges expected after suspected illegal fishing boat detected' (Sydney) 5 May 2001.

ABC News Online 'Federal Govt considers improving coastal surveillance' (Sydney) 5 May 2001.

The Age (Melbourne) 'Poachers loose in sail of the century' 6 May 2000.

The Saturday Mercury (Hobart) 'SAS hook fish pirates' 14 April 2001, 3.

The Age (Melbourne) 'Poachers lose in the sale of the century' 6 May 2001, 3.

The Weekend Australian (Melbourne) 'Illegal fishers caught after 4100km chase' 14-15 May 2001, 5.

The Australian (Melbourne) 'Sink poaching boats so they don't return- fishing boss' 19 February 2002.

The Age (Melbourne) 'Government urged to tighten fisheries defences' 20 February 2002, 4.

The Herald Sun (Melbourne) 'Fisheries call for armed monitors' 6 June 2002.

The Age (Melbourne) 'Fishermen may arm boats against pirates' 30 September 2002.

The Australian (Melbourne) 'Toothfish pirate uses SOS to escape' 8 October 2002.

The Weekend Australian (Melbourne) 'A Fishy Business' 26-27 October 2002, 28.

The Mercury (Hobart) 'Australia fails in Toothfish bid' 30 October 2002.

The Advertiser, (Adelaide) 'More illicitly profitable than the drug trade- Demand for this delicacy translates into $1million for each ship load' 23 August 2003, 38.

The Herald Sun (Melbourne) 'Fish poachers show colours' 25 August 2003, 15.

The Australian (Melbourne) 'Customs closes in on poachers' 28 August 2003, 3.

The Australian (Melbourne) 'Cold pursuit finally reels in Toothfish poachers' 29 August 2003, 3.

The Age (Melbourne), 'One rogue vessel captured but pirate king slips the net' 30 August 2003, 8.

AAP News Feed 'Pirates have organized crime syndicates' 30 October 2003.

AAP News Feed 'Pirates have organized crime syndicates:Toothfish operators' 30 October 2003.

The Mercury (Hobart) ' No bites on armed fisheries patrol' 13 May 2003.

The Mercury (Hobart) 'Snag in Toothfish War' 29 October 2003.

The Mercury (Hobart) 'Editorial' 8 November 2003.

The Sunday Age (Melbourne) 'Move on Fish Pirates Scuppered' 9 November 2003, 8.

The Mercury (Hobart) 'Letters to the Editor' 13 November 2003.

The Canberra Times (Canberra) 'Recalcitrants spike Toothfish watch plan' 19 November 2003, 23.

'Fishy Figures' (2001) The Economist.

10.2 International

Reuters 'Canada claims victory in fish wars' 16 March 1996.

Africa News Online ' South Africa Toothfish laundering' 20 April 1998.

BBC News Online (London) 'High-seas chase nets fish poachers' 18 April 2001.

The National Post (Canada) 'Australia chases suspected Toothfish pirates over icy seas: Poachers take major risks to catch fish valued as white gold' 19 August 2003, 13.

The Guardian (London) 'Southern Ocean hunt for ship with cargo of endangered Toothfish' 19 August 2003, 11.

Japan Economic News wire 'Australia pursues illegal fishers in longest maritime chase' 26 August 2003.

The Times (London) 'Toothfish pirates are boarded on high seas after 4,600 m chase' 28 August 2003, 11.

The Scotsman 'Fish poachers caught after Antarctic chase' 28 August 2003, 11.

Agence France Press 'Australia says capture of pirate trawler a warning to all poachers' 28 August 2003.

The Herald (Glasgow) ' Chase across oceans to net poachers, Viarsa finally boarded after being on the run for 20 days' 28 August 2003.

US New wire 'Statement by Manager of National Environmental Trust' 28 August 2003.

The Guardian (London) 'Antarctic chase ends with arrests' 28 August 2003, 19

The St John's Telegram (Newfoundland) 'An example for Canada' 30 August 2003, A10.

Africa News 'Fish poachers warned after Viarsa apprehended' 5 September 2003.

The South China Morning Post ' Fishing boat arrested in Australia after marathon chase, Crew of the Uruguayan vessel face fines, even jail terms for poaching Toothfish' 4 October 2003, 10.

The Nelson Mail (New Zealand) 'Law firm 'helping poachers'' 31 October 2003, 2.

11. Other Media Sources

ABC, Four Corners, 'The Toothfish Pirates' (broadcast 30 September 2002)

'White Gold' (ISOFISH)

The 7.30 Report , 19 August 2003.

ABC, AM Radio, 'Threat to Patagonian Toothfish' (broadcast 6 June 2002)

12. Personal Interviews

CSIRO, Dr Nick Bax (personal interview,15 November 2000)

Minister for Forestry, Fisheries and Conservation, the Hon Wilson Tuckey MP (phone interview, 19 May 2001)

Commonwealth Department of Public Prosecutions (Perth Office) (personal communication with Solicitors and Director: 11 May 2001, 1 June 2001, 22 June 2001, 29 June 2001, 28 August 2001, 26 November 2001, 14 February 2002, 23 October 2003)

Australian Fisheries Management Authority (personal communications with Mr John Davis and Mr Stuart Fitch: 14 June and 4 October 2001)

Australian Department of Foreign Affairs and Trade (personal communication with Dr Greg French, Director Sea Law, Environmental Law and Antarctic Policy Section 15 June 2001 and 4 October 2001)

Australian Department of Foreign Affairs and Trade, Sea Law (personal communication with Mr Michael Bliss, 15 June 2001)

Agriculture, Fisheries and Forestry- Australia, Fisheries and Aquaculture Branch (personal communication with Mr Kevin Bray, 14 June 2001)

Kalis and France Group, General Manager Policy and Environment (personal communication with Mr Martin Excel, 4 December 2001)

CCAMLR Secretariat (phone interview with Mr Esteban de Salas, Executive Director 20 December 2001)

CCAMLR Secretariat, (personal communications with Dr Denzil Miller, Executive Director 14 November 2003, 16 November 2004)

WWF, (personal communication with Margaret Moore, Senior Policy Officer – Marine, November 2003)

13. Research Papers

Fitzpatrick, J. and Newton, C. Assessment of the World's Fishing Fleet (May 1998).

Hanich, Q. Enforcement and Compliance Challenges to the Management of the Heard and McDonald Islands Fisheries Antarctic Cooperative Research Centre Research Report No.17, (2000)

Shearer, I. (ed) Prospects for a New Law of the Sea Martin Place Papers No. 2, International Law Association (Australian Branch) (1983)

Tsamenyi, M. and McIlgorm, A. International Environmental Instruments and their effect on the Fishing Industry Report of Fisheries Research and Development Corporation (FRDC) Project , University of Wollongong and AMC Search Ltd. (1999)

R. Churchill, The Meaning of the Genuine Link Requirement in relation to the Nationality of Ships Study prepared for the International Transport Workers Federation (2000)

Index

Reviews: Methods and Technologies in Fish Biology and Fisheries

1. J.R. Sibert and J.L. Nielsen (eds.): *Electronic Tagging and Tracking in Marine Fisheries.* 2001 ISBN 1-4020-0125-8
2. D. Symes and J. Phillipson (eds.): *Inshore Fisheries Management.* 2001 ISBN 1-4020-0128-2
3. C.O. Cunningham (ed.): *Molecular Diagnosis of Salmonid Diseases.* 2002 ISBN 1-4020-0506-7
4. T.S. Gray (ed.): *Participation in Fisheries Governance.* 2005 ISBN 1-4020-3777-5
5. R.J. Baird: *Aspects of Illegal, Unreported and Unregulated Fishing in the Southern Ocean.* 2006 ISBN 1-4020-5338-X